Praise for the work of Dr. Robert Melillo

"A visionary new approach . . . These valuable clinical insights add much to our tool kit."

—Daniel Goleman, author of *Emotional Intelligence*

"Clinically innovative and academically grounded . . . brings a refreshing, hopeful, and scientifically responsible approach to the field of childhood neurological disorders."

—Leslie Philipp Weiser, MPh, PhD, Harvard Medical School

"Dr. Melillo is a true visionary in the area of childhood developmental disorders, and autism in particular. I have met with children and families whose lives have been transformed by his work. I know that his latest book will be another useful tool that will help others learn about and possibly prevent autism in their children in a way that only he can."

—Zac Brown, Grammy Award–winning musician and founder of Camp Southern Ground

"Dr. Melillo's books are a ray of hope. Jam-packed with scientifically grounded information on brain function and its behavioral correlates, they provide satisfying explanations that parents recognize as relevant to their experience. And because his work is further tied to a simple intervention program, it has the power and potential to revolutionize the field."

—Michele Denize Strachan, MD, behavioral pediatrician, Developmental-Behavioral Pediatrics Program, University of Minnesota

"Dr. Robert Melillo is one of those rare individuals who can unravel the seemingly unsolvable mysteries of neurobehavioral development with wisdom, compassion, and vast perspective. His unique, groundbreaking, and research-based approach to improving brain function introduces the process of unlocking each child's potential."

—Pamela D. Garcy, PhD, clinical psychologist, professor of cognitive behavioral therapy, Argosy University, Dallas

continued . . .

Also by Dr. Robert Melillo

DISCONNECTED KIDS

RECONNECTED KIDS

AUTISM

The Scientific Truth About
Preventing, Diagnosing, and Treating
Autism Spectrum Disorders—and
What Parents Can Do Now

Dr. Robert Melillo

A PERIGEE BOOK

A PERIGEE BOOK
Published by the Penguin Group
Penguin Group (USA) Inc.
375 Hudson Street, New York, New York 10014, USA

Penguin Group (Canada), 90 Eglinton Avenue East, Suite 700, Toronto, Ontario M4P 2Y3,
Canada (a division of Pearson Penguin Canada Inc.) • Penguin Books Ltd., 80 Strand, London
WC2R 0RL, England • Penguin Ireland, 25 St. Stephen's Green, Dublin 2, Ireland (a division
of Penguin Books Ltd.) • Penguin Group (Australia), 707 Collins Street, Melbourne, Victoria
3008, Australia (a division of Pearson Australia Group Pty Ltd.) • Penguin Books India Pvt. Ltd.,
11 Community Centre, Panchsheel Park, New Delhi—110 017, India • Penguin Group (NZ),
67 Apollo Drive, Rosedale, Auckland 0632, New Zealand (a division of Pearson New Zealand
Ltd.) • Penguin Books, Rosebank Office Park, 181 Jan Smuts Avenue, Parktown North 2193,
South Africa • Penguin China, B7 Jaiming Center, 27 East Third Ring Road North,
Chaoyang District, Beijing 100020, China

Penguin Books Ltd., Registered Offices: 80 Strand, London WC2R 0RL, England

While the author has made every effort to provide accurate telephone numbers, Internet addresses,
and other contact information at the time of publication, neither the publisher nor the author
assumes any responsibility for errors, or for changes that occur after publication. Further, the
publisher does not have any control over and does not assume any responsibility for author or
third-party websites or their content.

AUTISM

First edition: January 2013

ISBN: 978-0-399-15953-4

An application to catalog this book has been submitted to the Library of Congress.

PRINTED IN THE UNITED STATES OF AMERICA

10   9   8   7   6   5   4   3   2   1

Neither the publisher nor the author is engaged in rendering professional advice or services to the
individual reader. The ideas, procedures, and suggestions contained in this book are not intended
as a substitute for consulting with your physician. All matters regarding your health require
medical supervision. Neither the author nor the publisher shall be liable or responsible for any loss
or damage allegedly arising from any information or suggestion in this book.

Most Perigee books are available at special quantity discounts for bulk purchases for sales
promotions, premiums, fund-raising, or educational use. Special books, or book excerpts, can also
be created to fit specific needs. For details, write: Special Markets, Penguin Group (USA) Inc.,
375 Hudson Street, New York, New York 10014.

CONTENTS

FOREWORD

Several years ago, I was lecturing at an autism society meeting about the role of environmental factors in autism. I remember that my first slide defined *autism* as a neuroimmune disorder induced by a combination of genes and many environmental triggers, including infections, toxic chemicals, and some dietary proteins and peptides. My last slide stressed the importance of the health of the parents before pregnancy, the health of the mother during pregnancy, and the infant's overall environment during the first year of life. At the time, this was considered a groundbreaking departure from the conventional wisdom that autism was a genetic disease, and for the most part, I felt somewhat alone in my understanding and purpose regarding autism.

The next day, however, I had the opportunity to listen to someone else speak very enthusiastically about the "disconnected kids" of his experience, and why our environment plays such an important role in the development of autism. And I knew that I was not alone.

That is how I came to meet Dr. Robert Melillo.

Dr. Melillo, the creator of the Brain Balance Program, is an internationally known practitioner, professor, researcher, lecturer, and author in the areas of neurology, rehabilitation, neuropsychology, and neurobehavioral disorders in children.

His drive and professional direction started fifteen years ago after seeing the life-altering changes that occurred within a child he had worked with. From that point, he has devoted himself to the study of learning disabilities and behavioral disorders in children.

On the day we met, I found out that the only difference between us was that my presentation at the time was based on blood work of autistic children and a review of the scientific literature, while his presentation was based on that *and* his direct or indirect clinical work with the thousands of autistic children who passed through his various brain centers, as well as his commitment to stopping what he calls the greatest childhood crisis in the history of mankind.

This kind of commitment requires a unique perspective, one that I call "the Melillo Perspective." You can only get this perspective from years of actual clinical experience of working with thousands of autistic children and their parents. Robert Melillo understands what is going on in the brains of these children, and he understands what is going *wrong* in the brains of these children. Without this kind of deep understanding of the problem, you cannot find a solution.

With his own understanding of the problem, Dr. Melillo became the architect of Brain Balance, a clinical program that can diminish, reverse, and even cure the social, behavioral, and clinical symptoms associated with autism and other neurological autoimmune conditions, and it is being practiced by thousands of young children right now across the United States.

So great is Dr. Melillo's commitment to fighting autism that he wrote his first book, *Disconnected Kids*, to serve as an at-home version of the Brain Balance Program for parents who could not afford to come to his clinics. He wrote his second book, *Reconnected Kids*, to help the parents of children with neurological conditions better understand their children's problems and work to diminish them.

The book in your hands is the logical progression of the previous books, and inarguably the most important to date. Whereas the other books proposed solutions and practices for dealing with autism and other neurological conditions, this book is about actually *stopping* autism, preventing it before it even has a chance to develop, and perhaps even curing it.

And that is why this latest book speaks to my heart.

Throughout the years, I have participated in presentations given by scientists who had been funded by NIH grants to conduct research on autism at many academic centers. With no exceptions, they have spoken about the genes associated with autism and ignored the role of environmental triggers. Most academicians appear to believe that autism is a genetic disorder that can be neither prevented nor cured.

This is in direct contradiction with the contents of this book and with the results seen daily in the thousands of children who are helped in Dr. Melillo's Brain Balance Centers, which is evidence that cannot be denied.

The contents of this book are comprehensive and specific. Chapter 1 details facts about the growing autism epidemic, facts that many academicians ignore, such as the fact that, yes, there is much we can do to prevent our children from having autism. Chapter 2 looks more closely at genetics versus lifestyle in autism and examines the role of epigenetics. Chapter 3 takes us into an autistic child's mind. Here Dr. Melillo says that certain environmental factors turn off the expression of key genes involved in building the brain, thus interfering with normal brain development. This is a statement that I agree with wholeheartedly; my own research shows that autistic children produce antibodies against their own neuronal cell antigens. Again, Dr. Melillo stresses that this means we can do a lot before and during pregnancy to prevent autism, once we know what to look out for. This leads to Chapter 4, which talks about the risk factors during pregnancy, exposure to chemical pollutants, breastfeeding, food sensitivities, and other dietary concerns of both mother and father.

It is the latter chapters of the book that deliver the actual payoff, because in these chapters, right here in your hands, are the secrets of Dr. Melillo's Brain Balance Program. Chapter 5 tells you how to determine your cognitive style, whether you are right- or left-brain dominant, and how this will affect the likelihood that your future child may have autism. Chapter 6 presents tests that will measure your brain imbalance. Most important, Chapters 7 to 9 tell you how to correct this imbalance, how environment, genes, stress, and your own body's immune system can be brought together, either crashing and clashing to bring disorder or working together in harmony to bring health to you *and* your children. And they tell you what to do to prevent disaster.

As the old saying goes, an ounce of prevention is worth a pound of cure. This is a view that both Dr. Robert Melillo and I subscribe to wholeheartedly. He concludes that, as with all diseases, prevention is the best cure. So he concludes in Chapter 10 by giving a ten-point plan for reducing the risk factors involved in having a child with autism, and the top ten lifestyle practices for a healthy pregnancy. Healthy minds, healthy bodies, healthy parents, healthy children. It's all in this book.

Millions of dollars and decades' worth of hours have been poured into research looking for the autism gene, the supposed magical key that will unlock the autism problem. But will it? We already know the genes associated with celiac disease, which is the best example of classic autoimmune disease that I can think of. How much has that really helped us with those disorders? We also know that gluten is the trigger, and much success has been achieved by removing these triggers through gluten-free diets and zero-tolerance practices.

The lessons to be learned from Dr. Melillo's scientific work can be summarized thusly: We are not powerless, and there is in fact much we can do to fight autism. Before it even has a chance to develop in our children, we can seize control of our own lives and do all we can to ensure their health. We can remove the triggers. For those who suffer, we can repair and improve functionality. We can change our lives, and the lives of our children.

I recommend this book for parents, parents-to-be, individuals with autism, pediatricians, doctors, nurses, and especially for academicians who believe that autism is ruled by genetics. If they learn only one thing from this book, may they learn how important in preventing disease are the periods before pregnancy, during pregnancy, and the child's first year of life.

For the past ten years, I have had the honor of knowing the eminent and talented Dr. Melillo. I have listened with pleasure to his lectures, and I have heard his message. And this is his message, to the parents of children with neurological disorders, to those who thought autism was a lost cause: There is hope.

There is hope, and you will find it in this book.

—Aristo Vojdani, PhD, MSc, CLS
Associate Professor, Technical Director, Immunosciences Lab Inc.,
and Chief Scientific Advisor, Cyrex Labs LLC

A Public Health Emergency

Disconnected kids, reconnected kids, functional disconnection syndrome (FDS), Peter Pan, and the Brain Balance Program are some of the issues and solutions that Robert Melillo has introduced in his earlier popular books to help parents of children with difficult problems.

Robert Melillo has used his clinical experience and research over time to develop assessment tools and a now-nationwide holistic program to help parents, particularly those of children with autism spectrum disorders (ASD).

But let us first talk a bit about the history. Autism was first described by the Jewish physician Leo Kanner (1894–1981), born in a small village in Galicia in Austria-Hungary. He studied medicine in Berlin, graduated in 1921, and emigrated to the United States in 1924 to take a position at a state hospital in South Dakota. In 1930, he was selected to develop the first child psychiatry service at Johns Hopkins Hospital, Baltimore, where in 1933 he became associate professor of psychiatry. He was in reality the first physician in the world identified as a child psychiatrist, the founder of the first academic child psychiatry department at Johns Hopkins University Hospital, and his first textbook, *Child Psychiatry*, published in 1935, was the first English-language textbook in this field. In his 1943 paper on autistic disturbances of affective contact, he described the specific syndrome of autism. He applied the Greek word *autos*, or self, to what he termed early infantile autism (Kanner's syndrome), characterized by the inability to relate to and interact with people from the beginning of life, the inability to communicate with others through language, an obsession with maintaining sameness and resisting change, a preoccupation with objects rather than people, and the occasional evidence of good potential for intelligence.

Interest in ASD has exploded in the past decade following the dramatic increase in the prevalence of ASD published in the most recent reports and discussed in this book. The increase in ASD cases has been accompanied by an increase in public and professional awareness and an abundance of new research and treatment strategies. What was once thought to be a rare, severe disorder is now recognized to be a common neurobehavioral disorder, which occurs along a broad continuum. Autism is probably not a single disorder, but rather reflects many different disorders with broad behavioral phenotypes causing atypical development.

Early identification of ASD and early provision of treatment can improve outcomes for many affected children, and this is an important aspect of the discussion in this book.

You do not have to agree with everything put forward in this book, but the interesting aspect is the focus on our environment and less on genetics, even though, as Dr. Melillo points out, many resources have been devoted to genetic research around ASD.

This book should be of interest not only to parents, but also to professionals working in this field, and it should stimulate more thinking around the mystery of autism and ASD that we have experienced in the past decade.

—Joav Merrick, MD, MMedSci, DMSc
Professor, Department of Pediatrics, University of Kentucky
College of Medicine, and Director, National Institute of
Child Health and Human Development, Jerusalem

Stopping the Greatest Childhood Crisis in History

The intention of this book is one of hope rather than despair. Optimism is not something you frequently find in a scientist who specializes in autism, and I am happy to be the exception rather than the rule. If more people could see autism from my perspective of hope, there would be fewer minds clouded with pessimism, though it is an understandable response to what is an all-too-common diagnosis.

I know autism and its impact intimately, as I have worked with thousands of children diagnosed with the neurological condition. I understand what is going on—and what is going wrong—in the brains of children as well as any expert and better than most. I come from a unique perspective of hope because I believe there are actions we can take, *beginning right now*, to stem this pervasive disorder that is threatening the mental health of our children. I know this for a fact, because I am already seeing it happen. I am the architect of a clinical program that can diminish, reverse, and even cure the social, behavioral, and academic symptoms associated with autism and other neurological conditions, including another even more common condition known as ADHD. The program is called Brain Balance, and it is being practiced by thousands of young children right now across the United States.

I am deeply entrenched in the science of autism and have written about it extensively in scientific papers and textbooks. For the last two decades I have had the privilege of lecturing about my research and experience with Brain Balance to distinguished audiences around the world, from parents and teachers to scientists and other learned brain experts. I have taught thousands of professionals in the United States, Europe, and Canada how to diagnose and treat childhood disabilities. I serve as faculty on the university level, including teaching a course in functional anatomy in a PhD program in neuropsychology. Each year I am invited to speak about autism at various medical and scientific conferences all over the world. In addition to these various commitments, I make sure I have time to speak to the people who need this information the most—the parents and teachers who are immersed in the daily lives of children with autism.

Curing autism and stopping its rapid rise is my lifelong ambition. In 2007, my partners and I opened our first of one hundred (and growing) Brain Balance Achievement Centers, where we work with children diagnosed with what are now referred to as autism spectrum disorders, as well as other disorders such as ADHD and dyslexia. We offer a specialized program that works to correct conditions caused by an imbalance in the growth of the left and right hemispheres of the brain through a series of motor, sensory, cognitive, and academic exercises, as well as nutritional and behavioral adaptations. Our results have been remarkable. The success we were seeing right from the start was the impetus for my first consumer book, *Disconnected Kids*, an at-home version of our clinical program targeted to parents who do not have access to or cannot afford to enroll their child in one of our achievement centers. *Disconnected Kids* and my second book, *Reconnected Kids*, have helped the parents of tens of thousands of children with neurological conditions better understand their child's problems and work to diminish them.

The book you are holding is by far the boldest, and undoubtedly will be my most controversial book, because it is about what we can do to *stop* autism, a condition that is widely believed to be genetic in nature and, therefore, can neither be prevented nor cured. In the following pages, I dispute conventional belief. There is evidence to show that autism, especially the most common type known as essential autism, is for the most part environmental in nature, meaning it stems from our modern style of living. As a result, I believe there is much we can do to prevent autism, and it starts even *before* a child is conceived. If there are genetic predispositions toward autism—and evidence suggests there surely are—then it likely is also possible to predict which parents are at risk of having a child with autism. Likewise, there are also actions we can take to help reduce this risk. Preventing autism, as you're about to discover, is a responsibility shared by both men and women.

Let me say right from the start that this is not just my personal theory, nor is it based solely on my twenty-five years of research, clinical experience, and knowledge. Though much of the book is theoretical, my conclusions are grounded in the best cutting-edge research findings from around the world. I believe that by addressing the currently known environmental causes and suspects, in addition to understanding the genetic risks, we can make a major impact toward arresting this disorder. We can significantly help lower the risk of having a child with autism and, therefore, reduce autism's skyrocketing rate.

More Talk Than Action

I see this book as just the beginning. Without question there is much more work that needs to be done. However, we don't have the luxury of waiting, because in my mind the wheels of change are moving much too slowly. I'm not talking about getting the word out about autism. It's everywhere. Organizations such as Autism Speaks, as well as others, have done an outstanding job of raising public awareness through media commercials, billboard messages, and well-organized and well-attended public fund-raising functions. I commend them for their tireless work. Thanks to them, the public is well aware that autism surrounds us. Unfortunately, the powers that be who control public and private funding don't walk the talk. Despite heightened public awareness, autism is grossly underfunded in the research arena. Research money coming from private donations and government grants has been relatively small. This comes as a surprise to many people, who naturally assume that awareness equals action. Unfortunately, this is not the case for autism. The number of children diagnosed with autism each year is greater than the number of children diagnosed with cancer, diabetes, and AIDS combined, the conditions that receive the lion's share of research money. It is also estimated that autism costs this country an estimated $126 billion a year, and that number is expected to skyrocket over the next decade. This figure does not take into account the loss of job productivity to family members and the financial toll it takes on families.

To put it in perspective, here are some examples of where research funding goes each year to fight childhood diseases: Leukemia, which affects 1 in 1,200 children, receives $277 million a year in funding. The childhood disorder muscular dystrophy, which affects 1 in 100,000, gets $162 million. Pediatric AIDS, which affects 1 in 300 children, receives $394 million, and juvenile diabetes, which affects 1 in 500, gets $156 million. However, autism, which affects 1 in 88 children, only receives $79 million in research funding. The National Institutes of Health earmarks approximately $35.6 billion annually for research, yet it allocates only 0.6 percent of these funds to autism.

As you can see, the actual money spent on autism research is low in relationship to prevalence when compared to other health problems. That is not to say that these other disorders should not be getting the money they are allocated. I feel they *all* should be getting *more* money. I'm just pointing out the little-known fact that autism isn't getting the scientific scrutiny it warrants.

POLITICS, AS USUAL

Then there is the matter of how autism research money is spent. For the last twenty-five years, almost 95 percent of autism research grants have gone to scientists and laboratories looking for a specific gene that causes autism. For the most part, these are the same people who scoff at the idea that autism is rising in epidemic proportions. To someone like me who knows autism so closely, this is mind-blowing. You'll understand why I feel this way when you read Chapter 1, but for now just let me say that the idea that the cause of autism is purely genetic is pure nonsense.

This is not just my opinion; it is the opinion of many experts and researchers who understand that a dramatic increase like we've been seeing can only point to the environment. Yes, genes play a role, and you'll be reading all about that in Chapter 2, but the environment plays a much bigger role. Devoting virtually all funds to seeking a genetic defect after coming up zero for twenty-five years, quite frankly, is cockeyed thinking and a waste of money. Actually, it is a shameful waste of money when you consider the way funding allocations are handled. It all comes down to politics and special interests, as usual.

Let me explain. Any scientist will tell you that genetic diseases do not increase at epidemic rates. So, if genetic researchers admit that autism is growing at an epidemic rate, they are also admitting that autism is not primarily a genetic problem. Poof! There goes their funding and possibly even their jobs.

Then there is the big gun—the pharmaceutical industry. Much of the money that goes into genetic research comes from pharmaceutical companies that want to corner the market on a drug to treat a genetic defect should one be discovered. They throw big bucks at researchers, and researchers and universities don't turn down grant money. That's understandable, as it is how the system works.

This doesn't mean that every researcher and every institution are solely looking out for themselves. Quite the contrary. They are passionate in their causes and they certainly have good intentions. I believe research money is being steered in the wrong direction because there is still so much confusion about what autism is from a neurological perspective. This is where I step in.

A Neurological Perspective

All professionals, parents, and teachers involved with autism are familiar with the symptoms and characteristics that define the diagnosis, yet truth be known, most don't understand what's happening in the brain of a child with autism. They aren't seeing and treating autism from a neurological perspective. In fact, the way modern medicine diagnoses and treats autism has changed very little since the condition was first identified seventy years ago.

Brain Balance broke new ground in the treatment of autism because it recognizes that autism, as well as other childhood neurological conditions, stems from what we call a functional disconnect, meaning the right and left hemispheres of the brain are out of balance. In autism, it's what causes what we see as an unevenness of skills—a child's ability to perform certain skills well, and in some instances, exceptionally well, while lagging or even failing in other skills. I've written about this extensively in my textbooks and in scientific papers. I introduced it to parents and the public in *Disconnected Kids* and again in *Reconnected Kids*. This type of imbalance is not caused by a defective gene but has its root in the environment. I'm not talking toxic chemicals, though they do play a role. By environment, I mean many of our modern lifestyle habits—actions that, for the most part, we can control.

Our Brain Balance Program depends heavily on searching for and addressing specific environmental factors that are exacerbating a child's autism. This is my expertise, which in itself puts me in the unique position to actually predict what environmental and genetic factors are coming into play that may potentially cause autism. By identifying and tackling these issues we can begin the real job of identifying the root cause or causes of a brain imbalance and get down to the business of slowing, and hopefully stopping, this epidemic. That's what this book is all about.

I am constantly traveling and lecturing all over the world to professionals, teachers, and parents. I think I do a pretty convincing job using research to show people that this increase in autism is real. Laying out the facts always leads to the question of *why?* Why are these problems increasing at such a rapid rate? What are the environmental factors that are causing this increase, and how do they cause autism? The conservative answer has always been we don't know *for certain.* At one time, this was a satisfactory reply, but not anymore. As the rate of autism rises, I also recognize a higher degree of anxiety in the audiences I address. People are no longer satisfied idly waiting for the big discovery or a cure written in stone. That could take another seventy years. They want to know what can be done *now.* I agree. The time *is* now. It's time we hop off the slow-moving wagon train of progress and get our motors running. I welcome the opportunity to take the driver's seat.

This book lays out the facts as we now know them. It defines autism as it should be understood—from a neurological perspective, based on what is happening in the brain. It examines our environment and lifestyle choices that, based on current evidence, we believe or suspect are contributing to the increased risk of autism. It looks at the controversy and facts concerning genetic mutations and genetic predispositions. And it recommends actions to help men and women grow healthier families and reduce their risk of having a child with autism.

I want to be candid about what you are about to read. I admit my ideas are aggressive, that I am pushing the envelope. It's very possible that some of the statements and assumptions I make will be found to be off or even downright wrong at some point down the road. That's the nature of all scientific scrutiny. However, I can say with confidence that the majority of what I posture will be found to be 100 percent correct. Whatever the eventual outcome, following the recommendations in this book is not for naught. I offer safe and sound advice that without doubt will benefit the health of all parents-to-be and their children.

Many naysayers will scoff at the idea of preventing autism. Others may think it is a bit premature, that we still don't have enough scientific proof to advise someone how they may be able to prevent such a baffling disorder. As a researcher myself, I completely understand skepticism. I candidly admit that I make some assumptions and draw some conclusions that are not yet fully supported by conclusive research; however, they are based on evidence and my deep, vast expertise. The minority of so-called experts who have criticized my work and my two previous books will do so again. I am ready to debate them at any time. I also recognize that some of what you are about to read will make some parents of children with autism uncomfortable because they may feel like I am blaming them for causing the disorder. I can assure you that this is not my intention. *No one is to blame.*

Since day one, Western medicine has been in the business of treating disease rather than preventing it. An attitude of prevention as the first line of defense is gaining in popularity, but like autism research, it too is moving much too slowly. I see this book as a means to accelerate the process. We can't wait. If environmental factors, as many experts believe and research suggests, are at the root of autism, then eliminating or modifying these factors will ultimately lead to burning out this disease. If these environmental effects burrow prenatally in a parent's biological chemistry or in the womb, then prevention has to start with early education and preventive tactics way before a couple tries to conceive. Understanding and identifying the genetic predispositions associated with autism are equally important in order for men and women to examine their own risk.

A Call for Action

The meteoric rise in autism is the most pressing social issue of our time. Many experts—and I am among them—believe we are facing the largest childhood epidemic in history. Just two decades ago, the likelihood of having a child with autism was 1 in 10,000. Today it is 1 in 88, and 1 in 54 are boys, and there is new and better evidence indicating the gap in total prevalence is closer to 1 in 40. Martha Herbert, MD, a noted autism researcher at Harvard Medical School and pediatric neurologist at Massachusetts General Hospital in Boston, calls the rise in autism a "public health emergency." With numbers like these, how can anyone not agree!

It's hard for me and like-minded experts like Dr. Herbert to comprehend, but we have colleagues who actually shrug off these numbers and attribute the rise in prevalence to better and broader diagnostic tools. They also say that, in general, we are seeing more diagnoses because autism doesn't carry the stigma it once did, so parents are more open to testing. Though true, this still cannot explain away these astonishing numbers. Sadly, this type of rhetoric only muddies the confusion that surrounds autism.

Our nation's politicians would like us to believe that the biggest threat to our financial security is health-care costs and entitlement programs such as Medicare and Medicaid. It's time to open our eyes wide and see that the real threat to our country's financial future lies in the health and well-being of our children and other generations to come. Their welfare is being threatened in a way we have never before seen. The epidemic rise in autism and related childhood disorders that we have been seeing over the past two decades is only the tip of the iceberg, as it is a problem that is going to accelerate if we don't do something to stop it.

This book contains important information and an agenda for all of us—parents of children with autism who want to have more children, men and women who want to become parents for the first time, grandparents and potential grandparents who want to influence a positive legacy, and all the rest of us who want to secure a safe and healthy environment for future generations. My hope is that this book will make a difference.

—Dr. Robert Melillo
January 2013

PART 1

SEARCHING FOR CAUSES

CHAPTER ONE

The Autism Epidemic

Controversy and Reality

Let's say you're a young woman or man with your whole life ahead of you. You're married or thinking about getting married, and you can envision a house filled with the delight of children. Parenthood is part of your dream, as it should be. After all, it's your natural next step in the evolution of life. However, you have friends with a child with autism or know people who know people touched by it. Your heart goes out to them, but it gives you pause. You can't help but wonder: What are the chances of autism seeping into your life when you have children?

Or perhaps you're in your late thirties or early forties and already have a child or children and you want more. You know all too well the challenges of parenthood in today's modern world and you believe you're handling them pretty well. You've heard and have read plenty about autism and you can only imagine what it must be like rearing a child with the disorder. You don't want to upset the contented life your family now enjoys, so you can't help but wonder: Could a second or third child be at risk?

Or maybe you're the prosperous parents of grown children and want more than anything for your children to be able to give their children the best opportunities in life, just like you gave them. The economy poses its own threat to their future, but on top of that you've heard about the rising rate of autism and the financial and emotional toll it can take on families. You've also heard some rumblings that it seems to target well-to-do communities. You can't help but wonder: What are the chances that your healthy, high-achieving kids could end up burdened with the problems that go with having a child or children with autism?

Today the prospect of parenthood brings unprecedented concerns that never even crossed the minds of parents in previous generations. Things like autism, childhood obesity, dwindling school budgets, guns in school, Amber Alerts, the pressures of getting into the "right" preschool, to name but a very few, are twenty-first-century phenomena that could not have even been imaged a generation or two ago. Understandably, they can fray a parent's nerves. However, out of these and all the other threats and challenges facing our children today, the one that concerns me the most is the rising rate of autism. I believe it is the single biggest threat to the future well-being of our children than anything that exists now or has come before it.

Autism is all around us. Question any eight-year-old and he will not only tell you what autism is but most likely will also say there are kids at school or in his class who have it. We're reminded of its existence virtually daily—in newspaper headlines, on the evening news, on highways billboards, at fund-raising events, or from a friend or neighbor who reports a recent diagnosis in the family.

Autism is a real threat. The Centers for Disease Control and Prevention (CDC) calls it a public health crisis and estimates that 1 out of every 88 children and 1 out of every 54 boys in the United States fits the diagnostic criteria. More recent and better research statistics indicate that the prevalence is probably even higher. This is a startling fact, for sure, when you consider that autism was a rare disorder just a generation ago. Startling also when you consider that, in terms of disease, autism is relatively new to the landscape of human life.

THE "BIRTH" OF AUTISM

Autism was first identified in 1943 by Leo Kanner, MD, a psychiatrist at Johns Hopkins University, after he encountered a five-year-old boy whom he described as being "withdrawn and living within himself." In his now-classic research paper published that same year in the journal *The Nervous Child*, Dr. Kanner wrote: "He seems to be self-satisfied. He has no apparent affection when petted. He does not observe the fact that anyone comes or goes, and never seems glad to see father or mother or any playmate."

The child, who Dr. Kanner identified as Donald T., was but one of eleven "problem children" he was seeing at Johns Hopkins with similar troubling personality and social characteristics. These children, he noted, "have come into the world with an innate inability to form the usual biologically provided affective contact with people." Hence, he called the condition autism for the Greek word *auto*, meaning self.

At the time he also noted, "These characteristics form a unique 'syndrome' not heretofore reported . . . which seem rare enough, yet probably are more frequent than is indicated by the paucity of observed cases."

How right he was. Unbeknownst to Dr. Kanner, around the same time across the Atlantic, similar observations were being documented by a German-born psychiatrist by the name of Hans Asperger, MD. In 1944, one year after Dr. Kanner's paper was published, Dr. Asperger wrote about a pattern of behavior in a group of young boys in Vienna that included lack of empathy, clumsiness, intense absorption in a special interest, and the inability to form friendships. He described these boys as "little professors" because of their fixation on a particular interest and their amazing ability to talk about it in detail uncommon to the average person, let alone a young child. He called the condition autistic psychopathy.

Asperger's and Kanner's path would never cross. Dr. Asperger's paper was published in German and got little notice until his research was translated into English more than forty years later. Shortly thereafter, a British autism researcher by the name of Lorna Wing, MD, coined the condition Asperger's syndrome. Hence, the close relationship and frequent confusion between autism and Asperger's.

THE STARTLING NEW STATISTICS

At the time of Dr. Kanner's discovery and for a decade or so thereafter, autism was considered a rare condition that was seldom seen or talked about. At the time, the incidence of autism was believed to be 1 to 2 children in 10,000. By the 1960s, the number had doubled to 4 in 10,000. In the 1990s, the incidence was even higher at 30 to 60 kids in 10,000. However, in 2005, research conducted at King's College in

THE AGE OF "REFRIGERATOR MOMS"

In the 1950s, autism took on a horrific taint when a man by the name of Bruno Bettelheim, who ran a school for disturbed children in Chicago, blamed the condition on emotionally distant moms. He called them "refrigerator mothers," an unfortunate and totally unfounded characterization that put a stigma on the condition and in many ways impeded efforts to recognize autism for what it really is—a neurological disorder caused partially by genetics and largely by our environment.

London concluded that the prevalence of autism was actually much higher than previously thought. Today records gathered by the CDC show that an estimated 1 out of 88 children in the United States, including 1 in 54 boys, are diagnosed with autism. This represents an increase of *600 to 1,000 percent* over the last two decades, depending on who's doing the math. Even at its most conservative estimate, this is an alarming increase—one that puts autism in the category of an epidemic. Polio was declared an epidemic when it affected 1 in 2,700 children.

These numbers, however, are highly debatable, largely due to the manner in which they traditionally have been calculated. For instance, the CDC collects its data by going through public school registries and counting the number of children who have been diagnosed with an autism spectrum disorder compared to the number of children who have not been diagnosed. It includes only children old enough to be in school and disabled enough to get a diagnosis or need services. This means that children not of formal school age and those with milder symptoms are overlooked. The CDC method of measuring autism prevalence is not unusual. Other research groups have collected autism statistics in a similar manner here and in other parts of the world and, generally, the results are similar to the CDC finding of 1 percent of the childhood population. In research circles, however, going through school registries is a far less accurate measure than surveys conducted in the general population. This point of contention came to a sudden head in 2011 when the first large autism population study was conducted—and it found the incidence to be an eye-opening 1 in 38.

The study, published in *The American Journal of Psychiatry*, was conducted by a joint investigative research team from the United States, Canada, and South Korea and involved 55,000 children in a small South Korean community. For the study, the researchers used standard autism diagnostic criteria to test children between the ages of seven and twelve who were enrolled in special education as well as mainstream schools. In addition to finding 1 in 38 children with an autism spectrum disorder, the researchers made an interesting revelation: Two-thirds of the children in mainstream school who fit the criteria for autism never had been diagnosed. The fact that autism, already considered to be rising in epidemic proportions, was actually even more prevalent than forecasted raised concerns and made headlines all over the world.

The study does not suggest that Koreans have more autism than any other population. Rates in developed countries around the world are generally similar. This pivotal study was telling us that autism most likely is universally more common *everywhere* than previously thought. Before this study was conducted, officials believed that the prevalence of autism in South Korea was around 1 percent of the population—similar to what we see in the United States. It turned out to be 2.6 percent of the population—three times higher than previously believed. Many experts—and I am one of them—believe that if we used the same study methods here in the United States, our numbers would be the same as South Korea's, or even worse.

The study moved Geraldine Dawson, PhD, the chief science officer for Autism Speaks, to proclaim, "These findings suggest that ASD is under-diagnosed and under-reported and that rigorous screening and comprehensive population studies may be necessary to produce accurate ASD prevalence estimates."

There is no doubt in anyone's mind that there are more people alive today with autism that at any time in history. It is indisputable that the numbers we are seeing today represent a dramatic rise in just the last ten years. The debate begins over why there are *so many* people today with autism. It's the proverbial $64,000 question: Does the rise in autism numbers represent a *true* epidemic increase in autism or not?

What's Behind the Statistics

The issue is largely one of semantics: *Prevalence*—that is, the number of individuals with autism—versus *incidence*, which means the number of actual new cases of autism. Prevalence, but not necessarily incidence, is on the rise, some argue, due to several factors:

- The term *autism spectrum disorder*, which has only been in use since 1987, includes autism; Asperger's syndrome, a high-function form of autism; and pervasive development disorder not otherwise specified (PDD-NOS), which is too mild to fit either profile. Prior to 1987, only those who fit the classic description of autism were figured into the count.

- Because the terminology has changed, the diagnostic criteria, which is based on the subjective answers to a series of questions, has been broadened.

- Terms such as *mental retardation* have pretty much become antiquated and frequently are now classified and diagnosed as autism.

- There is more diagnostic testing taking place due to an increased awareness of the disorder.

- The social stigma associated with autism is fading, so more parents are willing to have their children tested.

When it comes to laying claim that the rise in autism is epidemic, the distinction between prevalence and incidence becomes very important. Yes, there is no question that these factors have contributed to the higher prevalence we are seeing today. There is no question that including children with Asperger's and those classified with

A SUBJECTIVE DIAGNOSIS

Autism is quite different from most childhood conditions, for which taking a blood sample or doing an imaging test can determine a diagnosis. Autism roams the brain, not the veins, so there is no blood test. With autism, there is nothing organically wrong with the brain—it is not diseased or broken in any way—so there is no imaging test that can detect it the way, say, a mammogram reveals a suspicious tumor or a few suspicious cells. Rather, autism is a malfunction in the way the brain acts, so the only way to diagnose it is to observe a person's behavior. This makes a diagnosis highly subjective, based on the answers a parent gives to a series of questions and the observations of the evaluator.

The one constant in a diagnosis is the criteria used: the *Diagnostic and Statistical Manual of Mental Disorders*, developed by the American Psychiatric Association. Autism is diagnosed when a child exhibits a specific number of behaviors that signify deficits in three areas: communication, social interaction, and repetitive behavior. This may not be the best method, but it is the only method we have.

Subjective testing such as this is highly fallible. For example, we know that what are considered to be normal social skills vary significantly from child to child. The same goes for the various behaviors associated with autism. Many experts argue that most of the so-called symptoms of autism are really just extremes of normal behavior. For instance, at what point does a behavior cross the line from normal to abnormal? How much eye contact is considered "normal"? It's possible that evaluators can be either more generous or less scrupulous in their assumptions about a child's behavior, especially since a diagnosis of autism carries less stigma today than it did in the past. It's also possible for an evaluator to skew more toward a diagnosis, rather than away from it, depending on the mood of the moment. The subjective nature of diagnosing autism only adds more fuel to the debate over the rising rates of autism we are seeing today.

PDD-NOS have spiked the prevalence. And yes, there is more and better diagnostic testing taking place.

Even R. Richard Grinker, PhD, an anthropologist from George Washington University who was part of the Korean study, waffles over what percent of the study's findings can be due to better diagnosis and an expanded definition of autism versus the number of real new cases. Dr. Grinker, who is a specialist in South Korean studies, notes that back in the 1980s the stigma associated with autism was so great that Korean families generally were unwilling to admit that they had a child with a neurological problem. By 2011, he found that attitudes had largely changed. Nevertheless, he says he can't discount the fact that 1 in 38 makes autism way more common than we had thought. If we performed a population study in the United States in the same fashion conducted in Korea, he predicts the outcome would be about the same. However, Dr. Grinker hedges on calling these autism statistics epidemic. "You simply can't take prevalence estimates of autism as if they are the kind of hard scientific evidence that you would get from mapping out the increase in a virus," he told the press.

I have to say I agree with this statment. Estimating autism prevalence is a much less exact science, which is why the agrument has become so contentious.

ANOTHER POINT OF VIEW

There are also some researchers who believe that the prevalence of autism has always been high—that the rising rate is merely telling us that the condition has been previously misdiagnosed or underdiagnosed. Researchers in England attempted to prove this in 2011 with a survey of the adult population over the age of sixteen. If prevalence was on the increase, their theory goes, then there would be fewer adults with the condition than children. The researchers found what they were looking for. Their survey showed that the number of adults and children with autism was very close. "The lack of an association with age is consistent with there having been no increase in prevalance," they concluded.

LOOKING AT THE FACTS

Two recent major studies in California have come up with what I believe is proof that the rise in autism is real—evidence that should put the argument to rest. Both studies show that there is more going on than can be explained by heightened awareness and better and more diagnoses.

In one study researchers from Yale University examined nearly five million birth records and twenty thousand records from the California Department of Developmental Services and concluded that the rationale cited above can only explain away, at best, less than half of the increase in autism. They said around 25 percent of the rise can be attributed to what is described as "diagnostic accretion," meaning that children who would have been diagnosed with a condition such as mental retardation ten years ago are now diagnosed as both mentally retarded *and* autistic. They believe another 15 percent could be attributed to higher detection due to better awareness. Another 4 percent they attributed to "geographic clustering," which in a nutshell means a high diagnosis rate in a community naturally leads to more testing, which leads to even more diagnoses (though there are other reasons for clustering). This adds up to 44 percent. The researchers also figured that another 10 percent can be attributed to a higher risk of autism among children born to older parents, a recognized risk factor. Add it all up and you only get 54 percent. The researchers had no explanation for what is behind the other 46 percent increase.

No matter which end of the argument you're on, no one can deny that 46 percent is a *big* number. When you look at all the research and can only explain away roughly half of it to such things as increased awareness and better diagnostic practices, you can only conclude the other half represents an increase in incidence—that is, new cases. In the minds of many, including myself, this qualifies as epidemic.

Further proof that incidence is indeed on the rise comes from one of the most comprehensive autism studies conducted in the United States. Researchers from the University of California, Davis, found an *800 percent increase* in children diagnosed with autism over a sixteen-year span from 1990 through 2006. They concluded that changes in diagnostic criteria, the inclusion of milder cases, and an earlier age of diagnosis could account, at best, for about 50 percent of the increase. "Autism incidence in California shows no sign yet of plateauing," they concluded. "Wider awareness, greater motivation of parents to seek services as a result of expanding treatment options, and increased funding may each have contributed, but documentation or quantification of these effects is lacking."

"With no evidence of a leveling off, the possibility of a true increase in incidence deserves serious consideration," they added. "Whatever the final determination with regard to overlooked cases of autism in the past, the current occurrence of autism, a seriously disabling disorder in young children, at rates of greater than 30 per 10,000 individuals—and still rising in California—is a major public health and educational concern."

I believe, as many top researchers do, that these two California studies represent the reality we are facing today. We are seeing an increase in prevalence *and* incidence. I am not refuting that there are more children—in fact, a very large number of them—now being diagnosed with autism that possibly would have been overlooked were it not for heightened awareness, better diagnostic resources, and the other reasons already stated. But there are also just as many new cases of autism being recorded. Even if new cases account for only half, *400 percent is a big increase.* The Davis researchers call it "a major public health and educational concern." I couldn't agree more.

TIME FOR ACTION

In many ways, the idea that the dramatic rise in autism can totally be explained away by the reasons cited earlier is in itself very disturbing. This would mean that autism has been going unrecognized, undiagnosed, and untreated for decades. It means tens of thousands of children who should have been treated and helped have been ignored. However, I for one don't buy it and there are many other experts who agree. The fact that very credible research can't explain the rise of at least 50 percent of cases says that incidence—that is, new cases—is also on the rise. Unfortunately, it is unlikely that the debate over what's behind the increase is going to go away soon. As I pointed out in my introduction, it strikes too close to the heart of special interests. To me, continuing to argue over what's behind this rapid increase is only delaying the important job of finding a way to stop it. Thomas R. Insel, MD, director of the National Institute of Mental Health, has publicly stated it is time to get over our past sensitivities and move forward. "This whole idea of whether the prevalence is increasing is so contentious for autism, but not for asthma, type 1 diabetes, food allergies—lots of other areas where people kind of accept the fact that there are more kids affected," said Dr. Insel.

I'm in Dr. Insel's camp. It is time to stop stepping so delicately around autism, as if we fear harking back to the 1950s and blaming it all on bad mothering. We're far beyond that now. You can't argue with the fact that more kids are being diagnosed with autism today than at any time in history. When the gap narrows from 1 in 10,000 to 1 in 88 in seventy years, as the CDC estimates, or 1 in 38, as the Korean study suggests, it boils down to one thing: It means autism is an urgent problem than cannot be ignored.

There is plenty of evidence showing that we are seeing more autism for the same reasons we are seeing a lot more obesity, diabetes, asthma, and food allergies in our kids. For the most part, it is what scientists call environmental—that is, our modern lifestyle.

A No-Blame Zone

I've probably met more parents of autistic children than anyone else in the world. I work with them, counsel them, or give talks and workshops to them practically every day. I hear them agonize over whether they could have done something different to prevent autism in their children. I always assure them that they are not to blame.

In recent years, more and more research has surfaced identifying certain environmental factors that have the potential to affect the brain of a developing fetus in a way that can lead to autism. There is also evidence that these detrimental factors can be set in motion even before conception. However, many researchers are reluctant to discuss them publicly before the evidence is irrefutable, because they see them as potentially explosive and fear repeating past mistakes that put the blame on bad parenting. Memories of the horrendous claim made in the 1950s that emotionally cold "refrigerator mothers" made their children autistic still looms large. Nonetheless, the scientific community owes to the public, and parents in particular, the important and credible scientific data we now have.

I believe we are underestimating the intelligence of today's parents. The scenarios I used to open this chapter represent real concerns I hear when I interact with people in my travels. Parents want to know if they are doing something to contribute to the rise in autism, and I believe they want to know what they can do to reduce the risk. Really, who wouldn't want to know if there are outside factors, especially controllable ones, threatening their now and future children?

I recognize that responsibly presenting the facts and making recommendations is a delicate balancing act. To this end, I want to be up front and very clear about one thing: Nothing you are about to read in any way is intended to blame today's parents, their parenting skills, or their lifestyle for causing or contributing to autism. After all, how can someone be blamed for acting in a way or doing something that is potentially risky before the risks were even known? There is no way a parent of an autistic child could have known or predicted the evidence I am about to lay out. The intent is to help curtail preventable mistakes *going forward*, and to better understand and treat children with autism spectrum disorder now.

Let's start by getting a clear picture of what autism is, and isn't, and the role genetics play in the development of autism.

CHAPTER TWO

The Causes

Genetics Versus the Environment

Walk down the street and randomly ask anyone what causes autism, and most likely you'll hear this answer: *Autism is genetic. There is nothing you can do about it.*

The notion that autism is a purely inherited condition is ingrained in the minds of the majority of people, including many doctors and scientists. Autism was declared a genetic disorder after a handful of small studies beginning in 1977—incidentally, a time when diagnostic testing for autism was poor—found that if one identical twin had autism, there is about a 70 percent chance the other twin will develop it. Since that time, massive amounts of money have been spent on scores of scientific studies looking for a mutant gene that causes autism, and they've all come up empty.

Let's put science aside for the moment and just use common sense. Think about it. If autism is caused by a defective gene, then why are almost all parents of autistic kids not autistic themselves? Since the majority of people with autism never marry and have children, then why aren't we seeing a decrease rather than an increase in the disorder? Does it really make sense that more and better diagnoses can explain why seeing autism in schools is commonplace these days when it was almost unheard of just a generation ago? What does it mean when mentally healthy couples have multiple children with autism, something we are also seeing more often these days? And why are we seeing more autism in well-to-do communities?

Genetics alone cannot answer any of these questions. In fact, autism as a genetically caused condition is rare. A study published in the *American Journal of Medical Genetics* shows only 5 to 15 percent of autism is caused by a genetic mutation. We call this *syndromic* autism, and it mimics only some of the symptoms of autism. The other 85 to 95 percent, which is referred to as *essential* autism, is being driven by some outside influence. Yes, genes play a role, but what's going on around us is playing a greater role. The proof is found in the relatively new science of epigenetics, which tells us that *our environment, to a great degree, can manipulate how our genes behave.* Epigenetics has given us a whole new understanding of how autism evolves.

NATURE VERSUS NURTURE: WHAT TWINS ARE TELLING US

The idea of using twins to measure inheritability dates back to the nineteenth century, when an English scientist by the name of Francis Galton coined the phrase "nature versus nurture." Twins studies are important because they offer scientists a unique opportunity to untangle the influence of genes (nature) versus the environment (nurture) in understanding the evolution of disease. Because identical twins come from a single fertilized egg that splits in two, they share virtually the same genetic code—what we call DNA. Fraternal twins, who come from separate eggs, share the same genes as any typical

WHEN GENES ARE A PROBLEM

Though scientists have yet to find a defective gene that causes autism, there are certain neurological conditions caused by a genetic defect that produce autistic-like symptoms. These conditions are often referred to as syndromic, or extreme, autism, but I do not consider them to be true autism. Unlike autism, these conditions are rare, and the incidence is not changing. They also differ from autism in other ways:

- They are so severe as to be almost or totally disabling.
- They involve a global problem, rather than just affecting certain skills or certain areas of the brain.
- They affect girls and boys equally.

Truth be known, disease-generating mutations arise very infrequently and can take generations to affect a large population to the degree that it would be a noticable effect. Also, gene mutations are aberrant, so nature has a tendency to get rid of them. They get selected out and simply disappear. One of the few genetic disorders that fit this description and we hear the most about is fragile X syndrome. Children with fragile X have developmental delays of the whole brain, and not just one side as is typical of autism.

sibling—on average, they have about half their genes in common. Comparing identical twins to fraternal twins helps scientists quantify the extent to which genes and the environment affect our lives. If identical twins are more similar to each other with respect to an ailment than fraternal twins, the theory goes, then vulnerability must be rooted in heredity. That's how scientists in 1977 came to the conclusion that autism is a hereditary condition.

The study, which was first published in *The Journal of Child Psychology and Psychiatry* and has been written about thousands of times since, involved eleven identical and ten fraternal sets of twins in which one of the siblings had been diagnosed with autism. The researchers found the probability rate that the second twin would develop autism was significantly high in the identical twins and relatively low in the fraternal twins. The handful of studies that followed, which were just as small, found the likelihood of a second identical twin getting a diagnosis of autism ranged from around 60 percent to 95 percent. Despite the obvious limitations of these studies, the idea that autism is virtually 100 percent genetic has perpetuated.

It wasn't until July 2011 that a larger well-controlled study blew this theory out of the water. This study, which was conducted at the Stanford University School of Medicine on 192 pairs of identical and fraternal twins, provided convincing evidence that environmental factors are actually more at play than genetics in determining whether a child will develop autism. As expected, due to the results of previous studies, the researchers found that if one identical twin had autism, the other had a chance of developing it too. Curiously, they also observed something unexpected. They found the likelihood of twins developing autism was actually higher in the fraternal twins than in the identical twins, and genetics alone could not explain why. This could only mean the condition was also being influenced by the environment. Fraternal twins do not share identical DNA, but they do share a virtually identical environment. When the researchers did a statistical analysis they found environmental factors accounted for more than half of the autism risk. Genes were about 40 percent influential, but environmental factors were *almost 60 percent* responsible for the risk of autism.

This is one of the most compelling findings yet indicating that environment is playing a greater role in autism risk than genetics alone. Digging deeper into previous smaller studies hints at the same thing. The fact that both identical twins do not end up with autism in all cases tells us that something nongenetic is at play. For example, one of the earlier studies, conducted in Scandinavia in 1989, found a very high concordance rate of 91 percent among identical twins, but the researchers noted that virtually all of them were also exposed to maternal stress in the womb and tested high for stress levels after birth. (The stress response, as you will learn, is inextricably tied to autism.)

This twin study tells us that nature and nurture are not the only elemental forces at play in our biology. The third factor is epigenetics, a chemical reaction tied to neither nature nor nurture that influences how our genetic code is expressed. The process can strengthen or weaken genes, and it can also turn them on or turn them off. In some ways, epigenetics acts like a passageway between the environment and our genes; in other ways, it operates on its own to shape who we are.

THE ESSENTIAL ROLE OF EPIGENETICS

Epigenetics has revolutionized the way scientists look at human biology and chronic disease. Let's take another look at twins. Even though identical twins are born with the same DNA, they can become surprisingly different as they grow older. Many things about them are absolutely the same—their hair, eyes, facial features, skin type. This is their shared DNA, a genetic signature written in pen—what scientists call a genotype. As they grow they develop their own personalities, their personal likes and dislikes, and individual interests. They can even develop different diseases. This is their phenotype—their individuality, the sum of who they are. It's what scientists call an epigenetic effect. Epigenetics transforms cells from their original undifferentiated state during embryonic development into specific types of cells, orchestrating how our genotype develops into our phenotype. Epigenetics illustrates that the way genes are expressed early in life travels different paths over time, depending on life experiences and environmental exposure. Think of your DNA as a piano and your genes as the keyboard. Each key represents a gene responsible for a particular trait. In essence, epigenetics determines how and when each key will be played—the melody that portrays our life. It trumps the effect of nurture *and* nature.

The science of genetics has come a long way in the last hundred years. At one time, we thought our genes were our destiny with the same certainty that we know that brown-eyed parents will have brown-eyed children. We believed genes couldn't be changed, altered, or modified. This idea has led many people to believe that if heart disease runs in their family, then heart disease will strike them too. However, genetics is no longer thought of as defining our destiny as we once believed. Unless you have a rare genetic marker for one of the few diseases known to be caused by a defective gene, the way you live

IT'S THE ENVIRONMENT, STUPID

There is probably no one better qualified to speak about genetics and disease than Francis S. Collins, MD, director of the National Institutes of Health and architect of the Human Genome Project, which set out to identify the approximately thirty thousand genes in human DNA. In 2007, he told a U.S. Senate committee that the rise in autism is not genetic. Rather, he said, it is due to "changes in the environment."

"Genes alone do not tell the whole story," he wrote. "Recent increases in chronic diseases like diabetes, childhood asthma, obesity, or autism cannot be due to major shifts in the human gene pool as those changes take much more time to occur. They must be due to changes in the environment, including diet and physical activity, which may produce disease in genetically predisposed persons."

What stands out in this statement is that he did not mention anything about toxic chemicals, such as pesticides and heavy metals. He is solely referring to lifestyle.

your life steers your fate more than your genes. It's an epigenetic effect. However, unlike DNA, it can be altered—it is written in pencil.

Why Genes Don't Fit

The notion that genetic inheritance is the sole cause behind the rise in autism just doesn't hold water. The example I like to use is the obvious increase in the number of overweight and obese people we are seeing today. Like autism, statistics show obesity is rising in epidemic proportions in the United States and other places around the world. As a result, scientists have been pouring millions of dollars into research looking for a "fat gene," just as they are looking for a mutant gene that causes autism.

Again, let's use common sense. If obesity is genetic, how does it explain that overweight and obesity in the United States have risen a dramatic 400 percent in the last twenty years, so that nearly 75 percent of adults and 33 percent of children are overweight or obese? Claiming that doctors are simply diagnosing it more doesn't make sense. The fact that doctors are using a BMI chart rather than the scale to estimate body fat can't explain it either. It's ludicrous to think that over the last two decades people with an obesity gene suddenly have become attracted to one another, married, and passed on their fat genes to their offspring.

Let's bring it into perspective. Think of all the environmental factors that come into play. Eating too much and moving too little are the most obvious. There's also our sedentary dependence on computers, cars, wireless remotes, and all the other technology that makes us move less and burn fewer calories. Heck, we don't even get up to answer the phone; we just reach in our pockets!

Then there are the oversized portions being pushed in front of us and bottomless refills of calorie- and sugar-laden soft drinks. Fast-food restaurants have become the contemporary version of family dining. More egregious is our dependence on convenience and packaged foods containing trans fats and other artificial chemicals that our bodies can't even recognize, which confuse our body chemistry and cause us to release too much insulin and store fat rather than burn it. Believing there is such a thing as a fat gene is to say that people born with it are destined to get fat no matter what. It means that eating lean and healthy foods and exercising regularly would do no good. Of course, there's no logic in this kind of thinking. However, we do know that some of us have to struggle to maintain a healthy weight. There are those among us who have to watch every calorie and gram of fat, go to the gym religiously, and weigh in every day just to keep our weight in check. If not, we can easily see ourselves joining the ranks of overweight Americans. Why do some struggle, while others appear to skate through life as skinny as a rail? Because some of us have a *propensity* to gain weight. We have a *genetic predisposition* to obesity. That's quite different than saying you possess a mutant or defective gene that dictates that obesity is your destiny.

It's the same way with the genes involved in autism.

TENDENCY, NOT DESTINY

If your parents were overweight all your life and you have a tendency to gain weight and must always watch your diet, it means you have an overweight phenotype. Your parents probably had the same phenotype. You may not be overweight now because you work at preventing it, but it would be pretty easy to just let go and, *bam*, you'll fill out just like Mom.

So, let's say you fall in love and marry a skinny guy from a skinny family who all eat with abandon but never get fat. You're partnering with a skinny phenotype. Would your kids be skinny or fat? It depends on how they interact with their environment. If your kids hung out in pizza parlors, played video games instead of sports, had a maid to clean up their room, and never walked if they could catch a ride, then it's pretty certain they'll end up with a weight problem. Obesity is not necessarily your kids' destiny, but the door is wide open to make it happen. However, eating well, avoiding junk food, playing sports, and generally living a clean and active life would keep the genes— *your* genes—that predispose them to gain weight under lock and key.

Now let's say it doesn't work out too well with your phenotype opposite and you end up marrying another guy more like yourself. Even though you were both skinny when you met, the two of you have been living the good life and you've both put on quite a few pounds. Treat your kids to the same lifestyle the two of you have become accustomed to and you're setting them up to become an obesity statistic.

You get the picture. Obesity needn't be anyone's destiny if they choose to live in a healthy environment. Some people, however, need to struggle more than others because of their genes.

Predisposition to a health condition—whether it is obesity, heart disease, cancer, or autism—is an epigenetic effect. Just like obesity, a child can "inherit" a tendency toward autism from a parent or parents with an autism phenotype, but it is not destiny. It doesn't have to happen. There are ways to manipulate or reverse epigenetic effects, and that includes those involved in autism.

How Epigenetics "Spreads" Autism

During pregnancy, many changes must occur as cells become progressively specialized tissue that develop into the heart, kidneys, brain, and the other organs that define the human body. This process is called cell differentiation and involves a cascade of epigenetic programs.

You might remember from high school biology that genes reside on chromosomes like a string of pearls. Each location on a chromosome is called a locus. Sometime there is only one type of gene at a particular location, but more often there can be two or more. These different variations of the gene at the same location are called alleles. Think of them as different-colored pearls fighting for the same spot. We inherit two alleles at each locus, one from each parent. If the two allele types are identical, there's no conflict. They fit tightly in position and your dye is cast. If your parents each have brown eyes, for example, you'll have brown eyes. If the two alleles have equal influence, then you'll end up with a mixture of the two—perhaps hazel or a shade of green. In many cases, one allele is stronger than the other and will dominate. For example, brown is dominant, and blue is recessive.

The activity of a gene is called gene expression. Gene expression is controlled by gene regulation. At one time, genetic scientists believed

FOLIC ACID AND FETAL PROTECTION

Epigenetics is a relatively new concept, but mothers-to-be have been using it for nearly half a century to prevent a severe birth defect called spina bifida and other neural tube disorders. The epigenetic effect—one in which key neural genes are protected—is folic acid supplementation.

Folic acid is the supplemental version of folate, a B vitamin essential to healthy brain development. Most Americans don't eat enough fruits, vegetables, and whole grains, so we have a tendency to be folic acid deficient, which raises the risk a pregnant woman will give birth to a child with a birth defect. However, since cereals are fortified with folic acid and pregnant women take prenatal vitamins, spina bifida and similar disorders are now rare. This preventive measure is probably the first known application of nutritional epigenetics.

one gene (locus) corresponded to one trait. They also assumed that variations of genes (alleles) corresponded in some way to the difference in traits like blue or brown eyes or black or brown hair. This is what I refer to as "regular" gene regulation. It is fixed and short-lived. However, we now know that most traits aren't like hair and eyes, where all siblings are blue-eyed blonds. Most traits vary quantitatively, such as height, temperament, digestive capacity, and immune efficiency. How tall each child will be, for example, depends on the activity of many genes in addition to the influence of a variety of environmental factors. This is epigenetic gene regulation. It is random and ongoing.

Most of the time, most of your genes in most of your cells are turned off or are silent. They aren't doing anything. Activation occurs at the cellular level when certain chemicals attach to genes. Depending on the outside influence, these attachments can chemically change a gene. The DNA does not change, but the attachment epigenetically changes gene expression.

Divergent pathways of the same gene in siblings, for example, explain why one twin might be easygoing and the other uptight. The most powerful influence on gene expression comes from environmental sensory stimuli such as light, sound, taste, temperature, touch, and gravity, especially during infancy and early childhood when genes are busy building the brain. Gene expression is also influenced by exposure to toxic chemicals and diet, especially in the womb. Social interactions and interpersonal relationships become an increasingly powerful outside influence on gene expression throughout life.

THE PARENTAL FACTOR

Until recently, scientists thought that epigenetic attachments disappeared during the production of sperm cells and egg cells, which led to the belief that there is no such thing as an epigenetic inheritance.

The fertilized egg, they believed, begins development with a clean slate. We now know that this is not the case. Sometimes epigenetic attachments piggyback on genes as they are passed to future generations.

There are a number of mechanisms by which epigenetic gene regulation occurs. One key process called DNA methylation controls the strength of gene expression, making it stronger or weaker and turning it on or off—thus, manipulating gene expression. A gene that should be on is in the off position (or vice versa), changing the course of the genetic expression.

Methylation is the most common epigenetic tag that tends to stay attached to DNA, even after it replicates during cell division. The methylated DNA not only stays attached throughout the life of the cell, but it is transmitted to all other cells in its lineage. This is how epigenetic changes become inherited, not only from cell to cell but from parent to child and through many successive generations of children. In the past, scientists believed that the only way a trait could be passed on was from a mutant gene that physically altered the structure of DNA. So, if a particular trait or disease seems to run in families, it was automatically assumed that it was genetic.

Methylation essentially turns a gene off, and it can occur at any time during life. If a methylated gene were passed on to a child in the off position, then that gene would also be off in the child. This means that if some environmental exposure methylates a gene important to fetal or early childhood brain development in an adult prior to parenthood or preconception, it can be passed on to offspring. Even though it would have never affected the parent's brain development, the passed-on methylated gene will interfere with healthy brain development in his or her child, resulting in autism or some other similar neurological disorder. This explains how two people without autism can have a child who develops autism. It also explains why we are seeing more and more families with more than one child with autism. If one sibling has autism, the risk that other siblings will develop autism is around 10 to 20 percent.

A parent-to-be who is most vulnerable to passing on this epigenetic effect is one with an autistic phenotype. There are certain personality types, or cognitive styles, common in autism. The short list includes deficits in social skills, antisocial behavior, an inability to read body language, repetitive gestures or tics, an awkward gait, an odd head tilt, and unusual clumsiness—what I call an unevenness of skills. When exaggerated or severe, the culmination of these traits translates to autism. However, if you take special notice, you should be able to spot some of the same traits in a parent or both parents, though they may be ever so subtle. You might also spot them in siblings and other relatives. You'll likely notice that an autistic child who has an exaggerated ability to remember details, do complicated math calculations, or play music may also have relatives with the same natural abilities or interests, but on a much smaller scale.

A CASE FOR NEW CASES OF AUTISM

Epigenetics is not responsible for all cases of autism we are seeing today. Some genetic mutations arise spontaneously in children with no evidence of them in either parent. Evidence points to one source of these spontaneous mutations: the environment. Studies suggest that these genes, called *de novo* (meaning *new*) mutations are the source of autism that arises in families with no history of autistic tendencies. Scientists call this sporadic autism and it represents about 10 percent of cases, and they seem to be most common in older fathers. In one study, scientists identified twenty-one *de novo* gene changes in children with sporadic autism that were not found in their parents.

This is not a genetic scenario that can be corrected by genetic engineering or treatment. The best way to stop *de novo* mutations is to identify the environmental factors causing them.

REWIRING SILENT GENES

The brain is the most adaptable organ in the human body. We owe this to neuroplasticity, the brain's ability to physically and chemically change based on experience and what is happening inside and outside of our bodies. In essence, neuroplasticity rewires the brain—also an epigenetic process—and it can happen very, very quickly. This means that if tomorrow's parents find out today that they have environmental factors raising their odds of having a child with autism, they can start to take measures right now to reduce that risk.

Out of the estimated thirty thousand genes in the human genome, 85 percent of them relate to the brain and are instrumental in brain development. No one gene is causing the rise in autism we are seeing today. Each gene involved in brain development on its own has a relatively small impact on increasing the risk of autism. Autism develops from a confluence of genes involved in building social and cognitive skills that are either too weak to perform properly or are turned off so they can't perform at all.

Researchers from Harvard University demonstrated how genes that are turned off could be turned on again after isolating a group of genes involved in learning in children with autism. They found the genes were silent, as if the on and off switch was broken. When they compared these genes to neuroplasticity genes that change the brain, they found a strong overlap, much more than what would be expected by chance. It means that neuroplasticity can literally jump-start turned-off genes, and turned on genes can promote neuroplasticity. This research also explains why some autistic children improve after repetitive-type treatment interventions.

"Sometimes genes aren't completely inactive," noted Christopher A. Walsh, one of the researchers. "We know that intensive training or enriching of the environment in animal models has a way of turning genes on that would normally be silent." He added, "People think of genetic diseases as immutable and untreatable. Studies like ours and others give more hope we might not need to replace genes one by one, but find other ways of activating the genes that might be silent."

Or, if we can find what is turning off these genes, then we can prevent autism to begin with. How to go about it is the heart of this book. However, before you can understand how to prevent autism, you need an understanding of how a normally developing brain turns into an autistic brain.

CHAPTER THREE

What's Happening in the Brain of a Child with Autism

My crusade to stop the rise of autism involves spending hundreds of hours each year talking to parents and teachers who are in some way personally or professionally affected by autism or a similar disorder. I start each lecture with the same question: *Who can tell me what's happening in the brain of a child with autism, ADHD, dyslexia, or any other similar disabilities?* I am no longer surprised to see almost no one raise a hand. I then rephrase the question to make sure everyone understands: *Raise your hand if you have no idea what's happening in the brain of a child with autism.* All hands go up, and I see surprise on the faces of many as they look around the room and see that they are far from alone in their cluelessness.

Every professional I've come across who works with autism and every parent who has a child with autism can give a vivid description of *the symptoms* of autism, but they come up blank when I ask my question. This is really no surprise, as it follows the credo of how

Western medicine is practiced: *Treat the symptoms.* Fixing the cause is a concept that is too slow to take root in our society.

To me this is the biggest issue in autism. We are too fixated on treating the symptoms rather than fixing the cause. The fact that most people do not understand what is happening in the brain to cause the unevenness of social, communication, and cognitive skills that are the hallmark of autism contributes to the confusion that surrounds this confounding problem—and why the way doctors and educators treat children with autism has remained virtually unchanged since the condition was first identified seventy years ago.

One Universal Theory

Theories about the cause or causes of autism are not in short supply. As you've just read, genetics is a common one, but a mutant gene causing the type of autism that we see escalating in epidemic proportions is yet to be found, and I doubt it ever will be.

In addition, there are several theories that blame autism on insufficient or excessive brain chemistry involved in regulating the central nervous system, such as the mood-elevating compound serotonin and the reward-driven neurotransmitter dopamine. However, even though children with autism show irregular chemical flow in the brain, there is no strong evidence it is causing the problem. Most likely, it's the other way around: What's going on in the brain is messing up brain chemistry.

One of the most interesting theories to come along is called extreme maleness, based on the fact that autism is much more common in males than females, by a ratio of 5 to 1. According to Simon Baron-Cohen, PhD, the renowned autism researcher at the University of Cambridge who came up with the theory, abnormal levels of testosterone in the womb are creating brains with too many male characteristics. For example, he notes that the male brain is very good at systematizing, as are people with autism, and the female brain is better at empathizing, while the autistic brain is not. This, along with other features, convinced him that the extreme male brain is the cause of autism. It's true that testosterone levels can be high in pregnant women who have children who end up with autism, but in my opinion this theory has a lot of flaws and is too limited. Notably, there are other characteristics that typically are stronger in men than in women, such as coordination and gross motor skills, that people with autism, including males, do poorly with. Also, there are other characteristics that men typically do better than women, such as reading body language, but people with autism do them poorly.

Another school of thought associates autism with a malfunction in the immune system, due to the fact that children with autism generally have problems with allergies and food sensitivities. However, there is no good evidence to support the idea that an immune deficiency is causing autism. Here too it is more likely the converse: Autism is compromising the immune system.

We also hear about and fear certain toxins, particularly mercury. Though certain toxins are associated with increasing the risk of autism, there is no proof that they are the cause. For one thing, a toxic reaction would affect the whole brain because it would have to travel through the bloodstream to get there. However, autism affects only certain areas of the brain and usually doesn't affect them all evenly.

There are several other single-cause theories, including one claiming autism is the result of a vitamin B_{12} deficiency, an idea based solely on a finding that children with autism are frequently found to be low in the vitamin. However, there are many children without autism who are deficient in B_{12} too.

In my view, these theories all point in the wrong direction. They are focused on searching for a cause among what are actually effects.

They look at the proverbial trees rather than considering the forest, or the big picture—that is, what's going on in the brain. When we look inside the autistic brain, we can see that something abnormal is taking place. I, as well as many other experts, believe that certain environmental factors are turning off the expression of key genes involved in building the brain, which is interfering with normal development. As a result, certain areas of the brain, particularly on one side, are unable to communicate well with other areas of the brain and, in some children, are not communicating at all. This creates what we call functional disconnection syndrome (FDS), meaning certain areas and/ or one side of the brain are either growing too fast or too slow relative to other areas, causing the two sides of the brain to get out of balance, which disrupts the communication between these areas. It is the only theory that can explain the unevenness of skills that are the hallmark of autism. It is why a child may be great at math but find reading difficult. It is why all children with autism appear drawn into their own world and are challenged in the area of social skills. It is why children with autism commonly are fixated on repetitive tasks, such as counting, or repetitive movements, such as arm flapping. The most extreme example of this unevenness of skills is the rare individual who displays seemingly superhuman exceptional abilities in one area but is extremely impaired in others—what we call savant syndrome.

I believe most children with autism have a developmental problem affecting the functions primarily on the right side of the brain. It means that either the right brain is growing too slowly or the left brain is growing too fast, or both. In either case, it creates a right-brain deficit and the symptoms we typically recognize as autism. It is the theory that makes the most sense and the one universal theory that can explain all of the symptoms. I believe that if all the top experts in the world got in a room together to examine all the theories that currently exist, they'd all agree that functional disconnection syndrome is the one that needs to be addressed.

A Functional Disconnect

FDS not only describes what is going on in the autistic brain, it explains the symptoms we see in a long list of other neurological disorders, including ADHD, Tourette's syndrome, dyslexia, obsessive-compulsive disorder, and sensory processing disorder as well as others. All these conditions are the result of either a right- or left-brain deficit; the symptoms are different depending on the side of the brain and the área or areas affected. No matter what label you attach to it—autism, Asperger's, dyslexia, whatever—they all fall under the umbrella of FDS. It does not mean that the brain is diseased or damaged in any way, because it is not. It is just not developing as it should.

When you look at the brain of a young child with autism, it looks relatively normal, but the function is clearly abnormal. This can only mean that something happened or is happening during early brain development that is affecting growth of healthy connections between areas of the brain that are essential for fast and effective communication. As a result, some functions of the brain are very good, and in some cases, better than they should be for the child's age, while other functions are obviously delayed. This may be one of the ways a functional disconnect occurs. At first, the brain grows too fast, and then at some point it starts to lag behind. Imaging studies show that the brains of older teens and adults with autism are smaller than people of the same age with healthy brains.

A functional disconnect can occur at any time during brain development, even in the womb, but it generally remains undetected until obvious symptoms start to appear. You notice your child's head has an odd tilt. He may have a funny gait—cute, but odd. He covers his ears as if in pain when he hears you bang around pots and pans. You're waiting for those first words that should have come months ago. You notice he's oddly withdrawn and doesn't express much interest in others. The struggles you see in autism are almost all associated with impaired right-brain skills. Not *all* areas of the right hemisphere are affected equally. Some areas are affected more significantly than others and some are not affected at all. The differences can be subtle and differ from child to child.

Margaret Bauman, MD, a pioneer in the study and treatment of autism and an associate professor at Harvard Medical School, is one of the many experts who agree with this assessment. "The autistic brain appears structurally normal, both by direct inspection and by routine neuro-imaging techniques," she writes. "So, for the most part, the brain of a child with autism doesn't show any pathology. There is not anything obviously abnormal in the structure of the brain. It generally looks normal. Children with autism, although having generally normal brain structures, do show abnormal growth of the brain."

Although most children with autism have a normal head size at birth, she notes, the trajectory of head growth shows an unusual pattern of overgrowth during the preschool years. "Recent imaging studies now suggest an overgrowth occurs early in the postnatal period," she writes, "and is most marked between two and four and a half years. It is followed by a deceleration of brain growth in older children."

Many other experts and researchers support this theory, and there are numerous studies, including my own or those I collaborated on, that back it up. These are but a few:

- In 2012, researchers at Carnegie Mellon University compared eighteen individuals with high-functioning autism and eighteen individuals with the same IQ who were not autistic in a learning task that involved identifying liars among computer-animated avatars uttering the same sentence but with different facial and vocal expressions. The people with autism had difficulty recognizing the body language that should have tipped them off to the liars. The researchers found underconnectivity between the two sides of the brain in the individuals with autism compared to the others. "Underconnectivity in autism may constrain the ability of the brain to rapidly adapt during learning," the researchers reported in the journal *Cerebral Cortex*.

- Researchers from Carnegie Mellon, the University of California, San Diego, and the Weizmann Institute in Israel were the first to do a relatively large study of toddlers with autism. Their study, published in 2011 in the journal *Neuron*, found that problems with hemispheric synchronization could be detected in children as young as one year old. They also found that the specific window of time in which the brain gets out of sync defines the type of symptoms a child will exhibit.

- In 2010, *Cerebral Cortex* reported a study in which researchers used imaging scans to examine the brains of fifty-three males in late childhood and early adulthood with high-functioning autism and found the behavioral problems they exhibited were the result of a decrease in neural activity between connections in specific areas of the right and left hemispheres of the brain. They found that long-range connections, which are the more mature connections, had not formed. The authors even suggested that magnetic resonance imaging (MRI) scans might be a way to identify young children with autism even before major symptoms start to appear. They also noted something found by other researchers who have examined the brains of people with autism: The bridge between the two sides of the brain, called the corpus callosum, was smaller than in males of the same age with normal brains. Some researchers believe this defect contributes to the cause of autism. However, I believe this "defect" occurs as a result of autism.

- In 2010, I presented a paper to the European Society of Pediatric Research in which we compared brain scans that measured electrical activity in the brains of children with autism and normal children. We found that the electrical activity in the right hemisphere was significantly reduced relative to the left in the autistic children, but not in the others. There was also significantly less communication between the two hemispheres. We believe this was due to an electrical imbalance between the two hemispheres, which was a result of a functional disconnection.

- In 2009, Brazilian researchers performed an electroencephalography, a test that measures neural activity in the brain, on a group of boys with autism between the ages of six and fourteen and compared the results to those found in boys of the same age who did not have autism. They found abnormal connectivity between the two hemispheres of the brain in the autistic boys but not in the boys without autism, according to the journal *Clinical Neurophysiology*.

- In 2008, researchers from the Mediterranean Institute of Cognitive Neuroscience examined findings that "provide evidence that abnormal long-range connectivity between structures of the 'social brain' could explain the socio-emotional troubles that characterize the autistic pathology."

Brain Basics: A Brain in Sync

The idea of a functional relationship between the left and right sides of the brain is hardly new. In 1949, Canadian neuropsychologist Donald O. Hebb, PhD, famously concluded that "cells that fire together wire together" after he conducted research to figure out how the brain thinks and processes information. The adage means that brain cells that are activated, fired, and come to threshold at the same time will literally build physical connections to one another. In order for this to happen, timing is crucial, especially during early brain development. If neurons don't fire at precisely the right time, the connection doesn't happen and it is lost forever. This glitch can happen at any time, especially in the womb when early right-brain development is taking place. This is why pregnancy is such a risky time for autism to begin.

Even though brain cells develop very rapidly in the womb, a baby does not have much of a brain at birth, just enough to perform the basic functions of life, such as breathing. The basic structure and cells of the brain are built during early pregnancy, but the growth that builds these basic survival skills takes place toward the end of pregnancy. At birth, a baby has only about 25 percent of its brain, which essentially consists of the brain stem. The reason for this is obvious, as a baby's head barely fits through the birth canal. Once born, the brain grows rapidly and by age three, 90 percent of the adult brain will be formed.

By building the brain, I do not mean adding brain cells. More brain cells than a person will ever need are formed while in the womb, and about half of them will be lost during the first crucial years of brain development. Once a baby is born, neurons are directed by and interact with the genes that will control neural development, an introduction that is made by the various forms of environmental stimulation they will also interact with for the rest of life. Taste, temperature, touch, movement, light, sound, vibration, and gravity are crucial types of stimulation that a young brain requires. This is why a mother's touch and tenderness are so important. Parents are key players in stimulating brain growth.

There are two basic types of brain cells involved in brain growth. The majority of the work involves neurons, which comprise what is known as gray matter. The job of a neuron is to receive and transmit information. It does this by activating genes that stimulate the chain of events that make growth happen. The tip of each neuron contains a bunch of tentacles that act like an antenna to collect information from the surrounding environment through electrical and chemical signals. If the signal is strong enough, it goes to a neuron's central processing center called the nucleus. If the signal is very strong, it will filter to a cable-like structure called an axon, which releases chemicals that float across a very small gap, called a synapse, and make contact with receptors which rest on the surface of antennae-like structures called dendrites on a neighboring cell. These chemicals do one of two things: They excite the neighboring cell to send the signal farther down the line, or they inhibit the cell and turn off the signal. The complexity of the brain results from millions of these signals happening at the same time to integrate information that forms a new memory, experience, or understanding.

Brain growth also depends on the involvement of another type of cell called glial cells. These cells are fatty and have the important job of wrapping themselves around a neuron's axon. They act as a kind of insulation that allows impulses to travel faster down the length of the neuron.

The interaction of these cells is what gives the brain its power to grow. It is a process that enables neurons to increase in size and thickness. Like a muscle, each neuron grows thicker with use. However, key brain growth comes from actually increasing the number of functional connections formed from cell to cell, which spread like the roots of a tree. These connections are already starting to form by the end of pregnancy, but at birth they are still sparse, like a young plant that has newly taken root. By the time a child reaches age three, a vast jungle of trillions of connections has taken hold. Each brain cell ends up with about ten thousand connections to other brain cells. With one hundred billion brain cells in the average brain, this means connections will number in the trillions. It is a process that never stops.

This whole process of receiving and transmitting information is started by receptors that send information to the brain from outside stimulators, much like Morse code relays information. There are receptors for all kinds of stimulators—light, sound, vibration, touch, temperature, taste, smell, movement, pressure, pain, etc. These receptors convert the stimulation into a signal, which activates a neuron. The neuron takes the signal and sends it along at a certain frequency of electrical activity to site-specific areas of the brain for decoding. If the signal is familiar, the brain's memory center will respond. If the signal is unfamiliar, the brain will then learn and store it in memory for future use. The chain of cells involved in making these connections is called a network. Initially, as brain cells start making more connections, they reach out and connect with their close neighbors and coordinate a firing pattern for quick and efficient communication. They will "fire together" at the same time and then they will "wire together" by creating networks.

Immature connections are seen as short-range, or local, connections. This is what you see in babies; their ability to communicate is very limited. As the brain starts to grow in infancy and builds more and more connections and coordinates with larger groups of cells, longer-range connections start to develop in distant parts of the brain, eventually crossing over to the other side. Long-distance communication is taking place from one side of the brain to the other. Walking, for example, is a skill that requires left and right brain communication. As these cells mature and generate faster and faster impulses, they coordinate more and more cells to fire simultaneously and over longer and longer distances. As more and more cells are able to fire together at the same time, they start to break up and segregate into networks that all do a similar job. Each network can connect to millions or even billions of other cells and networks in the brain. This is what we see as maturity. As a toddler starts walking, she is slow, clumsy, and uncoordinated. This is because the brain cells that are driving the action as also slow, clumsy, and uncoordinated. As the brain cells strengthen through their connections with other brain cells and synchronize with networks on the opposite side of the brain, walking becomes more balanced and fluid. Maturing brain cells mature the child. This principle is at the core of neuroplasticity, which says that the brain has the ability to chemically and physically change throughout life, based on the stimulation and training it receives. Like protein and muscle fibers that build a bicep into a stronger arm through frequent exercise, brain cells will produce more connections and build a stronger brain through the type and frequency of stimulation it gets from the environment. It is how we continue to learn and remember new information throughout life. And it can happen rather quickly. Studies show that *within two hours* of learning something new, there are measurable differences in the size of the neurons in the brain.

It's a Matter of Timing

The brain is the most complex and important organ in the body, yet it's also the one we know the least about. One of the many mysteries of brain mechanics is how all these trillions of connections coordinate to form the big picture of our world. We do know that they are not all working together at the same time, as that would be impossible. The key appears to be the right connections happening at the right time, like a drawbridge that opens at a precise moment to allow a large-masted sailboat to pass through without holding up highway traffic.

Time is the glue. Time is what coordinates all of the cells to work together as one unit and gives the human brain all of its tremendous powers and ability. A memory or thought is made when cells from different areas of the brain all fire at the exact same moment. When this happens, a pattern of electrical activity or cells is frozen in time, making a memory or thought. To activate the memory means calling up the exact pattern of cells. If you can only recall part of a memory, it means you are only recalling a piece of the pattern. As you try harder to remember, more and more memory cells start to wake up. When they all come on line and synchronize, the pattern is whole and the "lightbulb" goes off in your brain.

Timing is one of the primary features that make the human brain unique in its intellect. It's what gives our brains tremendous processing power. An analogy I like to use to demonstrate this involves lining up ten people in front of a three-hundred-pound barbell. If I'd ask them to close their eyes, put one hand on the weight, and randomly try to lift the weight, they wouldn't be able to do it. However, if I said, "Count one, two, three, lift," they'd pick it up together, with no difficulty at all. The reason is timing. Time is the glue that makes lifting the barbell possible, and it is the glue that makes brain cells function as a whole brain. It is the difference between many cells firing independently or many cells firing together at the same time, essentially forming one big cell. The processing power is exponentially greater even though the number of cells is exactly the same. The human brain's ability to marshal cells and get networks to work together is what makes the human brain so superior.

The second thing that makes the human brain unique is lateralization—the phenomenon of left brain and right brain. Each side of the brain is designed to perform specific tasks that frequently must work in unison. For example, the left brain processes information, and the right brain interprets it. It means the left brain reads the words, but the right brain understands the story. We have dozens of specialized centers on either side of the brain that process particular types of information and control specific functions. The fact that we can then mix and match or combine these different areas of the brain within each side and between each side gives us an almost limitless repertoire of skills that we can tap into and develop. However, both sides of the brain must act simultaneously. The timing and speed of coordination gives the human brain its unique ability, but it also makes the brain vulnerable.

A Brain Out of Sync

Brain development is very sensitive to timing. It is as precise as the atomic clock. Everything is scheduled to happen in a precise order in a set window of time. For example, the terrible twos are the hallmark that brain development primarily has switched to the left side—from the negative right brain to the positive but independent and aggressive left. It's why hearing a toddler yell *no, no, no* is actually good, good, good. A child learns to crawl before he can walk. It's the path of normal brain development. If a child walks but never crawled, it is not a good sign. It means he literally missed a step in brain development.

The growing brain goes through many, many developmental windows of time before a child reaches each new milestone. Each window involves the expression of a cascade of genes that must come together at a precise moment in time. If some of these genes are turned off when they should be on, the window of time is missed and the learning doesn't takes place. It can happen in a millisecond or even a nanosecond. We can't see timing problems, and we can barely measure them. However, we know they occur, and I believe they are at the root of autism. As Dr. Hebb said, cells that fire together wire together. And if they don't fire together, they don't wire together.

This breakdown in coordination and timing in an otherwise healthy brain is the core principle of functional disconnection syndrome. In essence, a brain becomes autistic because environmental factors are causing epigenetic changes and/or spontaneous mutations to turn off, delete, or alter the genes that are primarily responsible for building connections in the brain. When interference in gene expression takes places during right-brain growth, it impacts the right-brain skill that should be taking place at that time. For example, the reason an autistic child cannot show compassion is because the genes that were supposed to build this skill misfired. The connection was never made. Physically, the glitch is so subtle it isn't obvious with a brain scan. In fact, a misfiring here and there might never have a noticeable effect on behavior or the ability to learn. In autism, however, it is more of a domino effect where one misfiring leads to another. If cells aren't sharing information and actively engaging one another, multiple connections are being lost. The effect is cumulative, especially during brain development.

In our laboratory, we have done research with autistic children in which we can see the imbalance in electrical activity and the way it is or isn't being shared. What we have found is a problem with "temporal coherence," meaning the timing and speed of processing between areas of the brain were measurably different, and the sharing and integration of activity was much less than is seen in a normal developing brain. We could see immature connectivity in each hemisphere of the brain compared to normal. We also noted that the coherence in the left brain was much better and faster than in the right side of the brain. This difference in electrical activity reflects a brain imbalance in which the right side of the brain is more immature and growing slower than the left. This is why children with autism typically have much better left-brain skills than right-brain skills—for example, excelling at math while having trouble with basic interpersonal skills. Why does it affect the right more than the left? Because both sides of the brain normally develop at different times, and the right hemispheres starts developing first in the womb and in the first two to three years of life. When an imbalance develops and is not corrected, it will get worse as time goes on. The two sides get out of sync. They become functionally disconnected.

Other researchers have found similar signs of immaturity in the autistic brain. For example, studies show that people with autism on average have an overproduction of short-range immature connections and an underproduction of the long-range, more mature connections, especially between the two sides of the brain. As I've already noted, studies also reveal that people with autism on average often have a smaller corpus callosum, the bridge that links the two sides of the brain. We are all born with most of the fibers in the corpus callosum, but it grows in size based on use—how often the bridge is crossed. As the two sides of the brain increase their coordination and integrate with each other, the fibers that make it possible for distant areas of the brain to communicate naturally become larger and thicker. A corpus callosum that isn't strengthening in size means neural traffic crossing the bridge is too light.

Sizing Up the Facts

Here's what we've learned so far:

- Certain genetic factors and predispositions, along with exposure to hostile environmental factors, before or during pregnancy and/or in early childhood development, combine to turn off or fail to turn on genes that build functional connections in the brain. This delays maturity of the brain.

- As a result, there is less coordination and communication among brain cells, which leads to fewer connections being formed than are necessary during a particular stage of development. When this happens in the womb or during the first two or three years, it impacts development of the right brain, creating an imbalance in the processing speed between the two halves of the brain. This imbalance causes the symptoms we know as autism. The more severe the imbalance, the more severe the symptoms.

- An immature right brain can actually be seen on tests that show cell activity in the brain. Low activity creates a processing imbalance, which means less information is being shared between the right and left hemispheres.

THE ANATOMY OF AN AUTISTIC BRAIN

Abnormal changes in the brain can originate before conception, during pregnancy, or in the early stages of childhood and can continue for the first several years of the child's life. Here is what we know about the anatomy of an autistic brain:

1. The brain seems to grow in an atypical way, but it is otherwise healthy, with no damage or disease. The cells themselves look normal, but certain areas may look more immature than others.
2. Although a number of specific brain regions are thought to be key areas of abnormality in autism, the disorder typically affects multiple areas, not just one area.
3. The brain regions primarily affected by autism are:

- *The limbic system*, which regulates emotions, behavior, long-term memory, and the sense of smell. In addition to the emotional and behavioral issues associated with autism, children with autism often display an underactive sense of smell.
- *The cerebellum*, which regulates certain cognitive functions such as language, attention, and motor skills. In addition to the language and attention abnormalities that are associated with autism, children with autism often have an odd gait, head tilt, exaggerated clumsiness, and/or repetitive movements.
- *The prefrontal cortex*, or frontal lobes, which houses what is known as executive function, an umbrella term for high cognitive skills, such as decision-making. Social behavior and personality expression are also housed here.
- *Brain stem*, which is structurally connected to the spinal cord and serves as a muscular and sensory pathway. It houses our sense of balance and gravity, which is called proprioception and is sometimes referred to as our sixth sense. Children with autism feel detached from their bodies and senses. (This is why I titled my first book *Disconnected Kids*.)

- Studies show that the brains of people with autism in general have more short-range (immature) connections and fewer long-range (mature) connections between the two sides of the brain.

- Research shows that the bridge between the two hemispheres, which is important for left- and right-brain communication, is smaller and less connected in people with autism.

- All of the symptoms of autism can be explained as either a left hemisphere that is too strong or a right hemisphere that is too weak, or a combination of the two. Because the brain controls all movement and body functions, every system is involved—immune, digestive, hormonal, and detoxification.

- This scenario does not explain *all* children with autistic symptoms. An estimated 15 percent have a clear-cut genetic mutation that affects males and females equally. The symptoms in these children, however, are generally more severe and are global, meaning they affect both sides of the brain. This type of autism is labeled *syndromic*. The other 85 percent have the condition just described. This is what we call *essential* autism.

- An estimated 10 percent of children with essential autism have genes associated with autism not inherited by either the mother or father. These mutations clearly are environmentally caused. Scientists call this *sporadic* autism.

- As it is with all disease, the best cure for autism is prevention. That is what I am going to address in the following pages. Let's start by looking at the various environmental factors that may contribute to producing an autistic brain.

CHAPTER FOUR

Risk Factors

What the Research Tells Us

Who is most likely to have a child with autism? Research tells us it is the person who is also most likely to succeed—a well-educated, highly intelligent male or female who excels at the hard sciences. In short, an engineer, scientist, or technology developer.

Evidence of this takes us to pockets of the world that are home to the highest percentage of high-technology firms, most notably California's Santa Clara Valley, better known as the Silicon Valley, and the headquarters to the giants of the industry such as Apple, Google, eBay, Yahoo!, and Cisco Systems. In all, the Silicon Valley is home to some sixty-five hundred high-tech companies, which have made this area one of the most affluent communities in the United States with a median salary of around $82,000, more than twice the median household income of the average full-time working American.

The Silicon Valley also has the dubious distinction of having an unusually high concentration of children with autism. Compared to the general population, where the prevalence of autism is conservatively considered to be 1 in 88, the rate of autism in the valley is 1 in 15. The prevalence of Asperger's syndrome, the high-functioning form of autism, is so endemic to the area that *Wired* magazine coined a cyber-age name for it, geek syndrome, in an article a few years back that asked the question: *Are math and science genes to blame?*

Well, the article never really answered the question, but I can. It is a qualified yes—and no. Yes, because there are a few good studies and lots of anecdotal evidence showing that a man or woman who possesses exceptional left-brained math and science skills can genetically predispose their children to autism. Two left-brained dominant parents raise the stakes even higher, which is what we're seeing in places like Santa Clara Valley. And no, because having a high risk or even the highest risk of having a child with autism does not mean it is inevitable. Risk can be reversed. As you learned in Chapter 2, a genetic predisposition is not the same as a genetic mutation. There is no such thing as a "science gene" racing around Northern California and other parts of the world where geek types reside. Being left-brain dominant is but one of approximately three dozen factors that can increase your risk of having a child with autism. However—*and this is important*—risk is cumulative. From what we can tell, it takes a convergence of risk factors to create the epigenetic effect that will either turn on or turn off a gene or group of genes that will ultimately interfere with the developing brain. The more risk factors you have, the greater your likelihood of having a child with autism.

My intention here is not to be the bearer of frightening news, and by no means am I attempting to talk anyone out of having children. I believe that today's brightest individuals give us the best chance as a society to raise children who will make the biggest differences in the world—our next generation of great minds who will create what is unimaginable today. Giving your children the best possible advantages, however, begins with giving them good health, and that starts with having two healthy parents.

It is important to remember that an infant brain is not a small version of an adult brain. The developing brain is much more susceptible to such things as toxic chemicals and other negative environmental factors than an adult brain is. This is especially true during early brain development in the womb. Because the brain grows in its complexity as a result of interacting with the environment, any negative impact early on can continue in a trajectory through various stages of brain development. The placenta provides a protective barrier against some environmental exposure, but not all of it.

When brain function is affected, other systems that are controlled or regulated by the brain are affected as well. Children with autism have problems with the digestive system, the immune system, and hormone production, and studies are finding that their parents, in many instances, do too. Most children with autism and many of their parents have a chronically high stress response and chronic inflammation in the body and brain.

Everything we know about the environmental factors that contribute to autism indicates that the health of the mother and father prior to pregnancy and the health of the mother during pregnancy are crucial. The environmental influences that can affect the way genes are passed on and expressed to your children can build up over a lifetime. It is possible that your mother or father passed on to you perfectly healthy genes that granted you normal brain development, but something you were exposed to while growing up—something as innocuous as hanging out on a buddy's farm or eating a particular diet, for example—could have turned off important brain-building genes, which you could pass on to your child in the same turned-off state. I explained in Chapter 2 how this can happen. Now I am going to explain the factors that can interact with an inherited predisposition to raise your risks, and what you can do to eliminate them. Virtually all these risks can be reversed or at least modified.

I've said this before and I am saying it again: In the end it all comes

RISKS THAT ARISE BEFORE CONCEPTION

The environment is responsible for genetic changes called *de novo* mutations that scientists believe occur in men and women prior to conception and are behind cases of autism without family heritage. Researchers from Harvard Medical School and Boston Children's Hospital have indentified the risk factors as:

- Paternal age of forty or older.
- Exposure to the heavy metals mercury, cadmium, nickel, trichloroethylene, and vinyl chloride and living in areas close to power-plant emissions.
- Vitamin D deficiency caused by decreased sun exposure, such as living in an area where there is a lot of rain or short daylight hours.
- Residing in urban areas located at high altitudes.
- People who spend a lot of time watching television rather than being active and getting outside.

down to a numbers game. There is no way of knowing now how many factors it takes to cross the line from risk to reality. It's doubtful we ever will. There are no guarantees. There are parents who will seemingly do all the right things and still have a child with autism and there will be parents who have lots of risk factors who will never have children with autism. For most prospective parents, it is a game of statistics, but one in which you are in control. Everything I am about to lay out is based on scientific research, not my or anybody else's hunch. I will tell you where the findings are the strongest, and the risks that are more speculative. Then, where appropriate, I will offer some bottom-line recommendations as to what you can do about them.

The impact of the environment on the development of healthy children is an important topic. The National Academy of Sciences started calling attention to it a decade ago when it publicly stated that environmental exposures are contributing to neurodevelopmental disorders in children. When the academy defined *environment*, it did not just mean toxic chemicals, but infections, nutrition, and lifestyle practices. In this chapter we are going to examine all of them, starting with the one that gets most of the attention these days.

ONE OR BOTH PARENTS WITH STRONG LEFT-BRAIN SKILLS

It is well documented that the personality type most likely to produce children with autism, especially a high-functioning form called Asperger's syndrome, are so-called geeks—those who succeed in the technology, science, and math arenas. These are people who are systemizers—they love precision, they are good at analyzing how things work, figuring out patterns, and breaking codes. They also tend to be introverted and uncomfortable in large social groups.

I vividly remember experiencing this firsthand a few years back when I gave a lecture to a group of doctoral students who were studying artificial intelligence at the Massachusetts Institute of Technology. They were incredibly brilliant at math and the sciences, as you would expect, but I couldn't help but notice that many of them were extremely clumsy and awkward in their movements. Most of them didn't make eye contact, and there was no humor or socializing going on while they were waiting for me to get started—not the kind of behavior you typically see in a university lecture hall. They barely even spoke to one another.

These students were oblivious to their physical and social oddities. They were extremely gifted in left-brain skills but deficient in their right-brain social and emotional skills. These students are the perfect example of a brain imbalance with left-brain dominance. They are not autistic, but they have some autistic-like qualities. One parent with this type of brain imbalance raises the risk for having a child who will have autism; two parents with this type of imbalance raises it even more. However, *how the genes of gifted people like these interact with the environment* can make the difference between having a child who simply inherits the same traits and one who ends up with autism.

Recent British research seeking an explanation as to why autism has an apparently high prevalence among families in the technology industry found that people with autism have a greater than average capacity for processing information, especially when viewed rapidly, and are better able to detect information defined as "critical"—the same thing we see among people working in the industry. The research, reported in the *Journal of Abnormal Psychology*, suggests that autism does not involve a distractibility deficit, but rather an information-processing advantage. This means that people with autism are not only good at certain skills, but they are better at them than most people.

The idea that people with a left-brain cognitive style are at risk for giving birth to children who will turn out to have autism has captured the interest of many scientists seeking clues to the cause of autism, including one of the world's most renowned autism researchers, Simon Baron-Cohen, PhD, from the University of Cambridge in England. He has done much research in this area already, and in early 2012 he started accumulating data online from parents around the world who are both scientists and comparing them to couples in which just one partner or neither are scientists. I believe his research will bear out what we already know. For example:

- When Dr. Baron-Cohen and his team from Cambridge conducted a study in Europe's Dutch version of the Silicon Valley they found an unusually high concentration of autism among the area's technology families. The researchers compared Eindhoven, the technology capital of the Netherlands and home to the electronic giant Philips, to two other towns where the demographics are the same except for the primary occupation of its inhabitants. They found the rate of autism in the towns of Haarlem and Utrecht to be the same as the general population, about 1 percent. In Eindhoven, it was twice that.

- A survey of undergraduate students, also conducted at Cambridge, found that those studying mathematics were more likely to be diagnosed with autism than were students majoring in medicine, law, or social sciences. The researchers also found that science and math majors scored higher in diagnostic criteria for having autistic traits than students studying the humanities or social sciences.

- Several studies have found that parents of children with autism, on average, have above-average IQs, especially on left-brain skills.

Studies also show autism is more prevalent in wealthier communities. Some argue that this is because awareness of autism is higher in well-to-do neighborhoods, where families have the means and are more willing to have their children tested. This can explain part of it, but not all of it. I believe it has more to do with left-brain intelligence. When individuals with strong left-hemisphere cognitive skills are given good educational opportunities, they tend to achieve great grades, go to better colleges, and end up meeting a like-minded spouse, who both get higher-paying jobs and live in the wealthier neighborhoods. They also have the highest risk of having children predisposed to autism, just like we are seeing in the Silicon Valley.

Recommendation: If you have a left-brain cognitive style, you probably intuitively know it, but you can confirm it by taking the Melillo Cognitive Style Assessment starting on page 132. This profile focuses on your natural strengths. You can also measure the severity of your brain imbalance by taking the Melillo Adult Hemispheric Checklist questionnaire starting on page 151. This assessment will identify your acquired weaknesses and the degree of the weakness. If corrective action is required, follow the advice in Chapters 7 through 9.

GENETIC TESTING "AT HOME"

Though there is no one gene that causes autism, approximately twenty genes involved in brain development have been identified that can contribute to autism. I am sure we will continue to identify more in the future. (You can find the names of these genes on page 264.) Scientists believe it takes a combination of environmental risk factors to alter genes in a way that will snowball into autism. If you have several of the risk factors discussed in this chapter—especially if you are left-brain dominant or come from a family of left-brained individuals—knowing your genetic makeup might be important if you are planning to conceive a child within the next year.

There are companies and institutions that do genetic screening, but it is costly and not practical for most people. Genetic screening is relatively new, and therefore uncommon, but my guess is that it will become routine sometime in the not-too-distant future. In the meantime, the operative word is *awareness*. One of the most cost-effective ways to spot a genetic link is by investigating your family history. For example, you can investigate if:

- You have family members or ancestors who were diagnosed with schizophrenia, suffered from anxiety, were socially deficient, displayed odd behavior, and/or displayed any other autistic traits. (A list of autistic traits can be found on page 127.)
- You come from a family of engineers and/or have relatives or ancestors who are unusually left-brain dominant and/or highly intelligent. This is of special note if it is common in both a father and a mother.

Undetected Brain Imbalance
in One or Both Parents

When it comes to a brain imbalance, the apple usually doesn't fall far from the tree. If you have a brain imbalance, it is likely your children will as well. If both parents have a brain imbalance, the likelihood that your child will have a brain imbalance gets higher. I believe that the single largest contributor to autism is *an undetected brain imbalance in one or both parents in which the right brain is too deficient and/or the left brain is too strong.*

There is no question that autism runs in families. This doesn't mean that people with autism are having autistic children. Rather, people with autistic *traits* are having children who end up with autism. This is how it can run in families. Studies have consistently found that the prevalence of autism among siblings is approximately fifteen to thirty times greater than the rate in the general population. Having one child with autism puts the risk of having a second child with autism at 1 in 500. If you have two children with autism, the risk of having a third is 1 in 3.

The more you lean toward the extreme end of left-brain dominance, the more likely it is that you have some autistic traits. They may be too subtle for anyone, including yourself, to notice unless you are looking for them. It's possible you had a brain imbalance as a child that corrected itself as you got older. Or perhaps you had a more subtle form of autism known as pervasive development disorder not otherwise specified (PDD-NOS) that was never diagnosed as a child. This risk factor alone may amount to very little in and of itself, but when you combine it with other environmental risk factors that you may have been exposed to over the years, it raises the potential of having a child with a more pronounced brain imbalance that could be diagnosed as autism.

If you display subtle autistic-like behaviors and traits, it is possible that you are carrying a genetic or epigenetic factor or epimutation passed on to you from previous generations that has gradually gotten stronger. For instance, your mother may have been exposed to a pesticide that did not affect her behavior or health, but did affect your development, though only slightly. As you were growing up, many of the environmental factors that could have exacerbated and magnified the problem did not exist. However, if they are affecting you now, they can affect your child.

The brain tends to get more unbalanced as we age and, for the most part, it tends to get more dominant on the left side. It is possible for you to have a different hemispheric imbalance than you had as a child, as the brain has a tendency to want to correct its own imbalance and, in doing so, may overshoot and create an opposite imbalance.

For the most part, the autistic traits found in one or both parents are subtle. John N. Constantino, MD, a researcher at Washington University School of Medicine in St. Louis, demonstrated that in 25 percent of families affected by autism, multiple family members exhibited autistic or subclinical autistic traits. Within some of these families these traits are extremely variable.

Numerous studies have documented the presence of autistic syndromes, symptoms, or traits, ranging from mild to severe, in the close relatives of children with autism. For example, a study in Denmark involving siblings of children with Asperger's syndrome found they had a thirteen times greater risk of developing full-blown autism than children in the general population. A study published in *Development and Psychopathy* tested 134 children with autism and their parents and found they shared a dysfunction in facial recognition. The only difference was the disability was much more pronounced in the children than their parents.

A brain imbalance has far-reaching biological implications with links to autism. For example, it is common for people with a right-brain deficit to have:

- A heightened stress response and associated high cortisol levels, which studies have found are common in children with autism, their mothers, and quite probably their fathers as well.

- Chronic inflammation and some of the health problems associated with it, such as high blood pressure, obesity, high cholesterol, and diabetes.

- A compromised immune system, which includes a greater susceptibility toward infections, allergies, and autoimmune disease. Studies found this is also common in children with autism and their mothers.

- Food sensitivities and digestive complaints related to leaky gut syndrome, which are common in children with autism.

- A tendency toward depression, anxiety, and other emotional issues. Socialization is a key deficit that defines autism.

Recommendation: You can find out if you have a strong left brain and an autistic phenotype by taking the Melillo Cognitive Style Assessment on page 132. You can further find out if you have a brain imbalance by taking the Melillo Adult Hemispheric Checklist questionnaire on page 151. If you find you have a brain imbalance, the exercises in Chapter 10 will help you get back in balance. Doing this before you conceive a child may help reduce your risk of having a child with autism or another neurological disorder.

OLDER MOMS AND DADS

One of the ironies of our fast-paced modern world is the phenomenon of the slower pace at which men and women choose to marry and have children. Unfortunately, in some way Mother Nature does not seem to approve. The optimum age for a woman to give birth with the fewest risks of complications and compromises to her own and her baby's health has not changed at all over the last several decades. It's around age twenty-four. However, the rate at which women beyond age forty are having babies has gone up an amazing 50 percent in just the last ten years, according to some estimates. Undoubtedly it is a trend that isn't going to reverse itself anytime soon. Studies show that the older you are when you conceive, the higher your risk of having a child with autism. This risk applies to both mothers and fathers, and the preponderance of evidence suggests the risk is greater for an older dad than an older mom.

A recent study conducted in Iceland and reported in the journal *Nature* found that the age at which a man fathers a child determines how many *de novo* (nongenetic) mutations will be passed on, increasing the chance that the child will develop autism or some other disorder. A woman is born with a lifetime of egg cells, but by contrast, sperm is continually being generated by dividing precursor cells, which acquire new mutations with each division. "The more mutations we pass on, the more likely that one of them is going to be deleterious," says Kari Stefansson, MD, chief executive of Decode Genetics in Reykjavik and the lead author of the study.

Dr. Stefansson and his researchers found that fathers passed on nearly four times as many new mutations as mothers and that the father's age accounted for nearly all the variation in the number of new mutations in the child's genome. They estimated that the number of new mutations being passed on rises exponentially with the age of the father. For example, a thirty-six-year-old man will pass on twice as many mutations to his child as a twenty-year-old. Though most mutations are harmless, the Icelandic researchers identified some that studies show have a direct link to autism.

This study is very revealing because it is telling us more than the fact that genetic mutations are more common in autistic children.

Rather, it shows that genetic mutations that arise spontaneously in older fathers are caused by exposure to some environmental factor or factors that accumulate with age and are most likely creating chronic inflammation. This makes it an environmental problem, not a genetic one. It also means it is most likely preventable.

When you take all the studies on paternal age and average them out, the risk of having a child with autism is 50 percent after age forty and rockets to 100 percent, or double the risk, after age fifty. One study suggests that older fatherhood could be one of the principle causes of autism in children in which there is no genetic link.

So, how old is old? Scientists draw the line at a woman's thirty-fifth birthday and a man's fortieth birthday. This risk is very real. Eleven studies that have looked into maternal age and autism found that the risk of having a child with autism goes up 50 percent after age thirty-five. For example, one study, published in the journal *Pediatrics*, showed that women who gave birth at thirty-five or older were 1.7 times more likely to have a child with autism when compared to women between the ages of twenty and thirty-four. Women thirty-five or older giving birth for the first time were nearly twice as likely to have a child who develops autism.

The risk is greatest of all for older men and older women who are first-time parents. One large study, reported in the *American Journal of Epidemiology*, that followed more than 250,000 children born to older parents found that the firstborns of a mother older than thirty-five and a father older than forty were three times or 300 percent more likely to develop autism than were third- or later-born offspring of younger moms and dads. The results caused the researchers to comment that "the increase in autism risk with both maternal and paternal age has potential implications for public health planning."

Why older parents, and older first-time parents in particular, are more likely to have a child with autism is still somewhat of a mystery.

Compounding the puzzle is another recent study, this one conducted in Denmark, involving more than 1.3 million children. It found that having an older mom *and* an older dad did not increase the risk more than having one *or* the other.

I believe *de novo* mutations are the best explanation found to date explaining the risk for older parents, especially fathers. However, I believe there are other, more subtle factors at play in changing our genetic profile as we age. Among them:

- It is possible that accumulated exposure to environmental toxins over a lifetime could result in genetic or epigenetic changes in the genes of older parents.

- Chronic inflammation and stress, which you will soon learn are intrinsically related to autism, are more likely to exist in older parents.

- As we age, our brains tend to get more out of balance. In fact, most adults tend to get more left-brain dominant with age, which is what we see in autism. Children with autism develop right-brain deficits, and I believe that if the parent has this type of imbalance, the parent can pass on the same imbalance in their children. The severity of that imbalance is based on how many other risk factors the parents have.

- First-time mothers are getting older and autism is more common in firstborn children. The average age for a woman to begin motherhood increased by 3.6 years, from 21.4 years in 1970 to 25.0 years in 2006. During the same time frame, first births to women age thirty-five years and over increased nearly eight times.

Recommendation: I do not believe that chronological age matters as much as the "real age" of your body—that is, your physical shape.

There are many ways in which you can determine your real age, which you can learn about on page 260.

A History of Psychiatric Illness

Studies have found a link between parents with a history of psychiatric illness and an increased risk of autism.

Most recently, a study conducted by researchers at Mount Sinai Hospital in New York found that schizophrenia was associated with a threefold increased risk of having a child with autism. There are similar findings for bipolar disorder but to a lesser degree. "These potentially shared risk factors could be genetic or environmental," the researchers commented.

Previously, a review published in the *American Journal of Epidemiology* ranked the disorders associated with autism in order of severity, beginning with the greatest risk, as follows:

- Schizophrenia-like psychosis

- Affective disorder

- Substance abuse

- Other mental disorders

An earlier study published in the *The Journal of Child Psychology and Psychiatry* concluded, "Parents of autistic children, especially fathers, were significantly more often rated as having schizoid traits."

Maternal Stress During Pregnancy

Scientists now know that fetal exposure to maternal stress can sometimes have deleterious effects, depending on the cause, timing, duration, and intensity of the mother's stress. A fetus can sense a mother's stress because stress hormones have the ability to pass through the placenta. Statistics show that mothers of children with autism reported significantly higher family discord, emotional problems, or other stressors during their pregnancies. For example:

- Four independent studies found associations between a mother's stress level around the time of pregnancy and the development of social and emotional problems during childhood.

- One study involving children between the ages of four and fifteen found an increased risk of autism among children exposed to their mother's stress in the womb, especially during the twenty-fifth to twenty-eighth weeks of pregnancy, the second trimester.

- Another study found that young children with autism who experienced prenatal stress between the twenty-first and thirty-second week of pregnancy were not speaking—a sign to the researchers that stress felt in the womb can be severe.

- One study that followed children born to 2,900 women showed that maternal stress during pregnancy in the form of typical stressful life events, such as divorce or a residential move, significantly correlated with the development of autism or ADHD by age two.

- One study found a correlation between continuing personal tension, specifically marital discord, in women throughout pregnancy and neurological problems, developmental delays, and behavior disturbance in their children.

- One study found that a mother-to-be's perception of negative life events, measured during weeks twenty-four and twenty-nine of pregnancy, was significantly related to increased risk of preterm birth, a risk factor for autism.

- A large study involving more than three thousand mothers who completed anxiety and depression questionnaires at approximately twenty-six weeks of pregnancy were predictive of premature birth and low birth weight, which are risk factors for autism and other health conditions.

- Work stress was found to be associated with an increased risk of premature labor, although it was most significant for women who experienced high job strain for at least thirty weeks of their pregnancy.

- Studies have found that children of mothers who report undergoing a separation or divorce during pregnancy or who experienced "cruelty by the partner" are more likely to display lower scores on tests that measure cognitive development, which is associated with autism.

As some of these studies indicate, the timing of the stressful event appears to plays an important role in the risk of having a child with autism. When researchers look at the brains of children with autism, they can see weaknesses in certain areas of the brain that are developmentally vulnerable at certain points in time during pregnancy.

There is a reason for this. The biological landscape for how stress is played out in the body changes during pregnancy. Cortisol, the primary hormone that is pumped through the body in response to stress, gradually increases two- to fourfold over the course of a normal pregnancy. This is necessary because cortisol plays a key role in the third trimester in helping the baby's lungs develop so he is able to take his first breath. It also assists in releasing the birth canal during delivery. Ideally, levels of cortisol should start low and gradually increase over the pregnancy, peaking at the end. As cortisol starts to rise during pregnancy, the placenta secretes an enzyme that acts as a partial barrier to inactivate rising cortisol, thus protecting the fetus. Toward the end of pregnancy, however, levels of this enzyme drop, allowing more maternal cortisol to reach the fetus in preparation for delivery. If physical stressors are causing the mother's body to release even more cortisol, it means the fetus is being exposed to more cortisol than is necessary or healthy. Hormone receptors in the central nervous system play a critical role in normal brain development, especially the regions that regulate emotions and cognitive function. Studies show that exposure to excess cortisol, particularly during these vulnerable periods of brain development, appear to be toxic.

In one study, researchers used saliva tests to measure the cortisol levels in 125 pregnant women. They also tested them on measures of anxiety and depression before, during, and after pregnancy. They then monitored the development of their children during their first year. They found that women who had higher levels of cortisol in the beginning of the pregnancy and lower levels at the end—the opposite of the way it is supposed to be—also had children who performed the worst on tests of childhood development. This may also be telling us that stress at *any* time during pregnancy is potentially detrimental.

Studies also show that the level of perceived stress by the mother and the actual response to stress in her body do not always correlate. This is because physical stress caused by anxiety or trauma is entirely different from the way the body handles its stress *response*. We can have *an increase in our stress response that is not in response to perceived stress.* This is key, because I believe that a chronically high stress response is possibly the single greatest contributor in autism. Lifestyle factors that affect our weight, diet, and activity level are key players in the level of our stress response. The relationship is so strong

---

THE TOP TEN STRESSORS

No two people respond to stress in the same way; however, research shows that these events typically produce the most stress:

1. Divorce
2. Marital separation
3. Death of a close family member
4. Injury or medical condition not affecting the fetus
5. Getting married
6. Getting fired from your job
7. Change in the health of a family member
8. A second or subsequent pregnancy
9. Business readjustment
10. Change in financial status

---

that I've devoted an entire chapter explaining how it happens and what you can do to literally turn it down.

Recommendation: The only way to know if you have a high stress response is to have it measured, which is relatively easy and inexpensive to do. You can find out about this test, and about the unhealthy and potentially risky cycle of increased stress response, in Chapter 8.

THE HEALTH OF A MOTHER-TO-BE

Pregnancy is a wondrous time in a woman's life, filled with the all the joy and expectation of waiting for the moment when she will hear the cry that will change her role in life forever. For the fetus, growing in the safety of the womb, it is a vulnerable time. What happens to the mom pretty much happens to her baby-to-be. Poor health and complications during pregnancy increase the odds that there will be complications at birth, which greatly increase the threat to a baby's health and neurological development. Any woman who wants to become a mom needs to be attuned to her environment and her body like never before—and this begins *before* pregnancy.

There is building evidence that the health of the mother is inextricably tied to the risk of having a baby who will develop autism. Researchers who examined the health of more than one thousand children born to mothers who were obese, had high blood pressure, or had diabetes prior to pregnancy found the risk of autism was 60 percent—*per condition*. Each condition increased the risk of having a child with a developmental delay by 150 percent.

It is becoming more and more clear that the months and perhaps even years leading up to the time of conception are critical to setting pregnancy on a healthy trajectory. In fact, many researchers feel preconception is the next new frontier in medicine. Of course, the risk of autism is only one of the implications a mother-to-be has to think about, but it is the only one I am dealing with here. These are the aspects of health in a mother-to-be that are linked to a increased risk of autism:

Overweight and Obesity

Americans are caught in a vicious cycle of increasing weight, with prospective mothers starting out heavier, gaining more weight during pregnancy, and giving birth to babies who are likely to become overweight adults. As would be expected in a "growing" population, this means there are more overweight women getting pregnant, which just recently has been linked to an increased risk of having a child with autism.

The study, reported in the journal *Pediatrics*, found that obese women are 67 percent more likely than normal-weight women to have a child who will develop autism. They also face twice the risk of having a child who will end up with other developmental delays. The study involved about one thousand children between the ages of two and four who were born to overweight women in California. Nearly seven hundred of them had been diagnosed with autism. The study defined obesity as being around thirty-five pounds overweight.

Estimates from the early 2000s show that the rate of overweight at the beginning of pregnancy jumped from 25 percent to 35 percent in just ten years. More disturbing, however, is the rise in pregnant women who are being classified as obese at the time of *delivery*. This rose 10 percent over the same period, from 29 percent to 39 percent. Though up-to-date statistics are not yet available, if the incidence keeps rising at the same rate, this would put the current rate of overweight and obesity at the start of pregnancy at around 45 percent, and around 49 percent at the end of the pregnancy. This means that *almost half* of pregnant women today are overweight and/or obese before, during, and after pregnancy.

It takes just basic common sense to know that this is not good. Overweight and obesity are the number one contributors to the rising rates of chronic illnesses plaguing Americans today, most notably metabolic syndrome, which is a cluster of symptoms that raise the risk for heart disease. These symptoms, which include overweight, high blood pressure, and diabetes, are implicitly connected to chronic inflammation, immune problems, hormone imbalances, and the high stress response that we are seeing in children with autism *and* their mothers. Children are not just inheriting a genetic disposition to become overweight from their overweight mothers, they are inheriting some of the major risks and problems inherent in autism.

There are a lot more high-birth-weight babies being born these days, and all indications suggest that it is related to the increase in overweight we're now seeing in the population in general. This is especially relevant because women are having babies at an older age, when they commonly weigh more than they did during their optimum childbearing years, the mid-twenties.

Even normal-weight women are gaining more weight during pregnancy than women were in the 1990s, according to data from the Centers for Disease Control and Prevention (CDC). Forty-odd years ago, a normal-weight woman was expected to gain no more than twenty-four pounds. It is not uncommon these days for a woman to gain more than forty pounds during a pregnancy with one child. In fact, more than a third of normal-weight women and more than half of overweight and obese women gain more weight than is recommended during pregnancy. One study, published in the British medical journal *The Lancet*, found that children born to women who gained more than fifty-three pounds during a full-term pregnancy were more than twice as likely to give birth to a baby weighing nine or more pounds than women who gained less than half as much weight. Nine pounds is considered the top end of a healthy birth weight. Because birth weight tends to predict body mass index later in life, "these findings suggest that excessive weight gain during pregnancy could raise the long-term risk of obesity-related disease in offspring," according to the authors. Sure enough, it is happening. Another study, also conducted in the United Kingdom, found that the children of women who were overweight during pregnancy were heavier than other children and were already accumulating risk factors for metabolic syndrome by age nine.

Women who are above their ideal weight at conception also run a higher risk of having complications during pregnancy and birth that are associated with an increased risk of autism. A study published in the journal *Circulation* found that overweight or obesity before pregnancy was more precarious than excessive weight gain during pregnancy in predicting a number of risks for the baby: birth complications and metabolic abnormalities associated with poor health outcomes, including childhood obesity.

Recommendation: If you're planning on getting pregnant, knowing your weight, as measured by the amount of body fat you are carrying around, is very important. You should know this prior to getting pregnant. The measure of body fat used by scientists is the body mass index (BMI). You can find out more about BMI and other measures of body fat on page 259.

Contrary to what you might believe, the heavier you are at the start of pregnancy, the less weight you should gain during pregnancy. These are the parameters for healthy weight gain during pregnancy:

- Thin women with a BMI of 18.5 or lower: twenty-eight to forty pounds.

- Normal-weight women with a BMI between 18.6 and 24.9: twenty-five to thirty-five pounds.

- Overweight women with a BMI between 25 and 29.9: fifteen to twenty-five pounds.

- Obese women with a BMI of 30 and higher: eleven to twenty pounds.

Smoking

Women who smoke during pregnancy are more likely to have a child who develops high-functioning autism such as Asperger's syndrome, according to a new study conducted by researchers at the CDC. The study compared smoking data from the birth certificates of thousands of children from eleven states who were diagnosed with autism.

The study doesn't conclude that smoking is a risk factor for autism, but it does say that there is an association between smoking and certain kinds of autism disorders.

Recommendation: Autism is only the latest in a string of health problems associated with smoking. The advice is pretty obvious: Don't smoke.

Autoimmune Disease

Researchers at John Hopkins University investigated more than six hundred thousand births over a ten-year period and found a link between autism and three autoimmune disorders in mothers. They concluded that if one of these disorders is present in the mother during pregnancy, the risk of autism in her child is increased as much as threefold. Autoimmune diseases found to have an association are:

- **Type 1 diabetes.** This is the insulin-dependent form of the disease in which the pancreas stops producing the hormone. It can strike children or adults at any age. Insulin sensitivity, though not necessarily diabetes, is commonly seen in autism. Other smaller studies found a link between type 1 diabetes in fathers and the risk of autism. The Johns Hopkins study found the risk to be nearly two times or 200 percent greater than in people who do not have the disease.

- **Celiac disease.** People with this condition, which has become more prevalent during the last decade, cannot tolerate gluten, a protein found in wheat, rye, and barley. Food sensitivities, including gluten, are common in children with autism. The risk of having a child who develops autism is 300 percent or three times higher in mothers who have celiac than those who do not.

- **Rheumatoid arthritis.** This is the crippling form of the disease that is more common in women and frequently attacks in early adulthood. The autism risk is 150 percent or one and a half times greater than in mothers who did not have the condition.

Recommendation: Autoimmune diseases have become much more common, and unless there are severe symptoms, these conditions are rarely detected. In fact, the most common cause of thyroid dysfunction, an autoimmune reaction called Hashimoto's disease, can often go undiagnosed for years. One of the most common causes of this disorder is a sensitivity to gluten, which almost always goes undetected as well. You can find out more about how you can test for autoimmune disorders in Chapter 9.

Fetal Exposure to Medications

Four medications or types of drugs are associated with an increased risk of autism:

1. **Thalidomide.** This drug, which was once used to treat morning sickness, made major headlines in the late 1950s for causing severe physical malformations when taken during pregnancy. However, the drug is still on the market and is used to treat Crohn's disease, an inflammatory bowel disorder, and multiple myeloma, a form of blood cancer.

   Studies show that taking the drug during the first trimester of pregnancy can increase the risk of autism. The most critical time of exposure appears to be twenty to twenty-four days after conception, a time that coincides with the closing of the neural tube and the start of nervous system formation.

2. **Misoprostol.** This drug is prescribed to prevent gastric ulcers. Studies show autism risk is correlated with taking the drug during the first trimester, approximately at the sixth week after conception.

3. **Valproic acid.** This is an anticonvulsant medication used to treat certain kinds of seizures and mania in people with bipolar disor-

der. It is also used to prevent migraine headaches. Ironically, it is sometimes used off label to treat outbursts of aggression in children with ADHD and autism. Children who are exposed to valproic acid in the womb are at risk for physical malformations similar to, but less severe than, what has been seen with thalidomide. Studies have found an association between taking valproic acid during pregnancy and having a child who displays autistic-like traits. The most critical time of exposure appears to be around the third and fourth week after conception.

4. **Antidepressants.** Emerging research suggests that prenatal exposure to the most widely prescribed kind of antidepressants called selective serotonin reuptake inhibitors (SSRIs) is associated with a modest increase in the risk of developing autism, especially when taken during the first trimester. One study found that mothers of children with autism were twice as likely to have taken prescription antidepressants in the year prior to delivery. The study, however, could not determine if the effect was tied to the medication or the mother's depression itself. SSRIs include the brand names Celexa, Lexapro, Prozac, Paxil, and Zoloft.

Recommendation: You noticed the critical time of fetal exposure to these drugs occurs early in pregnancy when it is possible you may not even know you are pregnant. If for some reason you are taking one of these drugs and want to become pregnant, talk to your doctor about getting weaned off it far enough in advance, so that the risk to your unborn child is not a factor.

Rubella Infection

Also known as German measles, rubella has been quite rare in the United States ever since a vaccine against it came into use in 1969.

Rubella is generally a mild disease, but coming down with it during early pregnancy is very serious, as it can cause miscarriage and major birth defects, including deafness and blindness. There is also evidence linking it to autism. According to a study reported in the *Annals of Neurology*, babies born to mothers who contract rubella during pregnancy have a 13 percent chance of developing autism. The riskiest time for autism is when the infection occurs during the first eight weeks of pregnancy.

Recommendation: The measles, mumps, and rubella (MMR) inoculation is the vaccine that will protect you from infection. However, you should receive it well in advance of conception and not during pregnancy. If you had the vaccine during childhood or had German measles as a child, you are not at risk for getting it in adulthood. If you have not had the vaccine, talk to your doctor about the best time to get it prior to pregnancy.

The Flu and Other Infections

As with rubella, it is not infections per se, but the timing of an infection during pregnancy that increases the risk of autism. It appears that the first and second trimesters are the most vulnerable times for contracting other viral infections as well as bacterial infections.

One landmark study, published in the *Journal of Autism and Developmental Disorders*, followed the health of children in Denmark born to mothers who came down with an infection during pregnancy over the span of fifteen years. The researchers wanted to know whether children whose mothers had an infection during pregnancy were more likely than their peers to develop an autism spectrum disorder. In addition, they explored whether the nature of the infection—viral versus bacterial—or the trimester during which the infection occurred affected risk. The results: Viral infection during the first trimester increased the child's risk of developing autism by 200 percent.

Bacterial infection during the second trimester increased the child's risk of developing autism by 42 percent.

The riskiest type of infection? The flu. Fifty percent of the children in this study who ended up with an autism disorder were born to mothers who came down with the flu during pregnancy. The rate of autism among children of mothers who were sick enough to be hospitalized for the flu was six times or 600 percent higher than in the general population.

There is some speculation that the flu itself is not harmful to the fetus, but rather it is the mother's inflammatory response that is the culprit. I tend to agree. You will find out how a compromised immune response puts your baby at risk later in this chapter.

Recommendation: Though there appears to be a link between the flu and autism, there is no link between getting a flu shot and the risk of autism. Women who are pregnant during the flu season or who expect to become pregnant should get a flu shot. This is something you should discuss with your doctor or obstetrician.

Fever During Pregnancy

Fever during pregnancy from any cause can increase the risk of having a child who develops autism, according to research conducted by the University of California's MIND (Medical Investigation of Neurodevelopmental Disorders) Institute. The study also found that the risk can be eliminated by taking fever-reducing medication.

The study involved more than eleven hundred children involved in the Childhood Autism Risks from Genetics and the Environment (CHARGE) study. The researchers found that mothers of children who were diagnosed with an autism spectrum disorder were two and a half times more likely to have had a fever during pregnancy than mothers of children who did not have autism. The risks were particularly elevated when mothers reported having had a fever during the second trimester of pregnancy. However, mothers who took over-the-counter anti-fever medication had the same risk of having a child with autism as women who reported having no fever during pregnancy.

The researchers suspect that that natural chemicals released by the body to fight a fever may pass through the placenta and interfere with brain development.

Recommendation: Any sickness during pregnancy requires the attention of your obstetrician. Contact your doctor right away. If you have a fever, take over-the-counter anti-fever medication, such as acetaminophen or ibuprofen, as directed by your doctor.

Gestational Diabetes

No studies have been conducted looking for a direct link between autism and contracting diabetes during pregnancy, but studies have found a link between the condition and other neurological diseases associated with a brain imbalance.

A recent study, published in the *Archives of Pediatrics & Adolescent Medicine*, suggests that babies born to mothers with pregnancy-related diabetes have a higher risk of developing ADHD, which is a right-brain deficiency just like autism. Another study, conducted at New York's Mount Sinai School of Medicine, found that children born to mothers who experienced gestational diabetes had lower scores on tests measuring language and memory skills and IQ at age three and four than children of mothers who did not have diabetes during pregnancy. By age six, these same children were scoring lower in communication and attention testing. The researchers suspect that extra glucose passes through the placenta and alters oxygen and iron levels in the blood, which affects certain areas of brain development. This means it is possible that gestational diabetes could increase the risk of autism.

One of the biggest risk factors for contracting gestational diabetes is being overweight at the start of pregnancy or gaining too much weight when pregnant. During pregnancy, a woman's body naturally lowers its sensitivity to insulin in order to shunt vital nutrients to the growing fetus. The body uses insulin to remove sugar from the blood and move it into cells for energy. However, both overweight and/or excessive weight gain during pregnancy exaggerates this normal process by further increasing insulin resistance and possibly altering other maternal hormones that regulate fetal nutrition. Excess nutrition stimulates fetal insulin secretion and fat production. It also puts women at risk for gestational diabetes.

Three things can elevate blood sugar in a pregnant woman: poor eating habits, lack of exercise, and an increase in the stress response.

Recommendation: The best way to reduce this risk is to modify your diet, exercise regularly, monitor your stress response, and rule out a brain imbalance before becoming pregnant. How to go about this is explained in the rest of this book.

Preeclampsia

Preeclampsia is high blood pressure that develops in the late second trimester or third trimester of pregnancy. It is considered a potential risky complication. Though findings have been mixed, research involving the mothers of more than one thousand children enrolled in the CHARGE study found that having preeclampsia increases the risk of having a child with autism by 60 percent.

To me, the link makes perfect sense. Preelampsia creates a metabolic disturbance in the body and is a sign of inflammation and a heightened stress response that can affect the developing brain of the fetus.

Recommendation: If you have high blood pressure going into pregnancy or have any risk factors for preeclampsia, such as overweight or obesity, you should be under the care of an obstetrician who specializes in high-risk pregnancies.

Compromised Ability to Absorb Folate

Preventing folate deficiency is what taking prenatal vitamins is all about. Folate is essential to early brain development, and folate deficiency in pregnant women is a well-known risk factor for the birth defect spina bifida and other neurological conditions. There is also increasing evidence that a deficiency in this B vitamin might increase the risk of autism.

One recent study showed that taking prenatal vitamins three months before or during the first month of pregnancy lowers a woman's risk of having a child with autism by 60 percent.

Another recent study surveyed the mothers of more than seven hundred preschoolers about their diet and supplement use before and during pregnancy and calculated how much daily folate the women were getting each month. They found that getting at least six hundred micrograms of folic acid, the supplemental form of the vitamin, was tied to a 38 percent lower chance of having a child with autism.

The risk, however, is greater than simply remembering to take your vitamins. Scientists have isolated at least two genes that can interfere with the body's ability to absorb folate and cause a vitamin deficiency. Having one of the genes can put a mother at risk.

This was demonstrated in another CHARGE study that looked at three groups of children between the ages of two and five whose mothers took prenatal vitamins. Two groups of mothers carried one of the genes, but a third carried neither of the genes. The researchers found that mothers with the high-risk gene called MTHFR had a 4.5 times or 450 percent greater chance of having a child with autism and mothers who had the high-risk gene called COMT had a 7 times or 700 percent greater risk of having a child with autism than mothers who did not have a folate-robbing gene.

The fact that all the mothers with the gene did not have children who developed autism is an example of how several factors—a genetic predisposition and certain environmental influences—combine to create autism.

Recommendation: I believe that taking prenatal vitamins can help reduce the risk of having a child with autism, just as it reduces the risk of having a child with a birth defect. It is a given that pregnant women should take prenatal vitamins, but there is no way of knowing if you possess a gene that blocks folate metabolism without testing. You may possess the gene if you have a family history of getting heart disease at a young age or unusually high cholesterol levels run in your family. I believe and research shows that folic acid supplementation should start even before a woman becomes pregnant, as it can compensate for an increased risk of autism, even if certain genes are present. This is something to discuss with your doctor. Though it is not clear if a man's genetic risk is the same, a man with these genes can pass risk on to his children.

Maternal Bleeding

A review published in the journal *Epidemiology* found bleeding during pregnancy was a "significant" factor in having a child who develops autism. The Harvard researchers speculate this is because bleeding during pregnancy is a sign of fetal hypoxia, or oxygen deprivation. "Fetal hypoxia may underlie a potential relationship between gestational bleeding and autism," they said.

Bleeding during the second half of pregnancy can be indicative of an injury to the placenta.

Extreme Morning Sickness

An extreme form of morning sickness known as hyperemesis gravidarum (HG) can take a toll on expectant mothers, causing dehydration and often requiring hospitalization. It is a form of forced starvation during a time of fetal brain development, when proper nutrition is paramount. It also induces stress in the mother-to-be and heightens the stress response.

Studies show that children of mothers who experienced HG beyond the first trimester have more attention and learning problems than children who were not exposed to HG in the womb. Though none of the studies specifically looked at autism, it goes without saying that anything that causes prolonged malnutrition and dehydration during fetal brain development is potentially harmful. In addition, HG has also been found to heighten the stress response in pregnant women. Any woman who experiences unusual and prolonged morning sickness should get prompt attention from her obstetrician.

TRAUMATIC BIRTH

One of the most dangerous trips we ever make is the journey through the birth canal. The fact that we walk on two feet, rather than four, has gained us much selective advantage over other life, except perhaps for that first trip of life. When humankind achieved the ability to walk on two feet millions of years ago, it required a repositioning of the pelvic bones, which resulted in a smaller birth canal. This posed a new challenge, because human babies have relatively large heads, even though the brain still has most of its growing to do once baby greets the world.

Obstetric conditions have long been associated with a number of neurological and neurodevelopmental disorders including mental retardation, schizophrenia, speech and language problems, internalizing problems, attention problems, social problems, and hyperactivity. It is widely recognized in scientific literature that the birth process itself very often results in injury to the baby, especially the cervical spine. Unfortunately, a cervical injury can go unnoticed by physicians who are unfamiliar with how to diagnose and/or treat it.

When left untreated, a cervical injury can result in persistent developmental and functional neurological problems as a child grows. Birth trauma that results in cervical injury can lead to a decrease in motion of the spine and cervical muscles that provide a large amount of the ongoing stimulus the brain needs to grow. The amount of sensory input from the neck is more important for healthy brain development than the sensory input from all other areas of the body combined. An altered mobility of the neck on one side or the other can cause a significant imbalance in the sensory input to the brain from day one. Loss of or an imbalance in this input is of significant concern because it can be a primary cause of a brain imbalance. Because a newborn's brain growth is primarily focused on the right side, this means it can lead to a right-brain growth deficit, possibly setting a child up for autism.

Many children with autism have motor (large muscle) problems from the day they are born, and a problematic delivery could be responsible. Considering the high percentage of children who probably suffer from this type of trauma, it may actually be one of the largest causes of the higher-functioning cognitive and behavioral problems we are seeing today.

Numerous studies over the last few years have claimed a link between autism and specific complications at birth. Researchers from Harvard School of Public Health and Brown University reviewed forty of these studies that explored fifty different birth-related conditions suspected of increasing the risk of autism. They confirmed some and exonerated others. Reassuringly, they concluded that no one birth complication in and of itself led to a high risk of autism. "Rather it appears that increased risk is associated with a combination of several factors that may reflect what is referred to as a suboptimal birth," commented Geraldine Dawson, PhD, chief science officer for Autism Speaks. "And even then the risk may only be present when combined with a genetic vulnerability."

The Harvard study as well as other research suggests a risk between these birth events and a risk for autism:

Premature Birth

Premature birth can be caused by any of a variety of complications during pregnancy, but for the most part it is a mystery. Doctors can't find a logical explanation for half of them. A normal pregnancy should last thirty-eight weeks, or forty weeks if you try to factor in the actual date of conception, but a baby is considered to be at full term at thirty-seven weeks. Any time earlier is premature.

Premature birth is associated with a variety of health problems, including developmental delays and later intellectual impairments in childhood and adolescents. A small number of studies have associated an increased risk of autism with birth prior to thirty-seven weeks. Some of them say that the shorter the pregnancy, the higher the risk. One study found that birth before thirty-one weeks was associated with a 700 percent increase in autism.

Low Birth Weight

Low birth weight, which is defined as less than five pounds, eight ounces, is considered a marker for putting newborns at high risk for later neurological and psychiatric problems with a laundry list of symptoms common to autism, including attention, social, and learning problems. Although a link between low birth weight and ADHD is well documented, there are only limited studies associating low birth weight with an increased risk of autism. One study involving eighty-seven thousand births, conducted by researchers at the University of Southern California, "significantly associated" low birth weight with greater odds for autism. Two other studies, plus the Harvard study, also found a clear relationship.

Logically, it makes sense. Because right-hemisphere growth begins in the womb, anything that can affect the growth and development of a fetus has a greater likelihood of causing a right-hemisphere delay.

There are a number of problems associated with low birth weight or being born small. Problems with the placenta can reduce blood and nutrients getting to the fetus. Maternal nutritional problems during pregnancy, as well as infection in the fetus, have been found to cause abnormal growth. Infection in the mother, diabetes, high blood pressure, and placental abnormalities are associated with growth restriction and autism.

Low Apgar Score

Within the first few minutes of birth, a brief neurological test is performed to assess the health of the newborn. Though named after Virginia Apgar, MD, the doctor who invented the process in 1952, Apgar is considered an acronym for the evaluation: appearance, pulse, grimace, activity, and respiration. The score ranges from zero to ten, with a seven to ten considered normal. An Apgar score below seven is associated with an increased risk of autism. A low Apgar score is more common in premature births and is generally an indicator of oxygen deprivation.

Breech and Caesarean Births

Studies show that children with autism are twice as likely to have been born by breech birth. Breech births are associated with increased stress to both the baby and the mother. Usually this will also lead to a more uncomfortable delivery with more pain and anxiety.

Breech birth also limits the movement of the baby in the uterus.

Babies should move and turn frequently, an important aspect of brain development.

There are many factors associated with having a caesarean section that could affect the mother and the child, including the use of anesthesia and other medications. Caesarean births also can increase the stress response in the mother.

Babies who do not experience vaginal birth miss out on their first opportunity to use their primitive reflexes, because they do not travel through the birth canal. Primitive reflexes allow a newborn to perform the basics of life, such as breathing and sucking. Developing and later shedding primitive reflexes is important to healthy development of motor skills. I discuss this more in Chapter 12, where you will find some diagnostic tools and therapeutic suggestions.

Small for Gestational Age

Several studies have found a significant relationship between a fetus being small for gestational age and autism. A number of factors come into play. Problems with the placenta can reduce blood flow and nutrients to the fetus. Maternal nutritional problems during pregnancy as well as infections in the fetus are associated with abnormal growth. Maternal infection during pregnancy, diabetes, and high blood pressure can also affect the nutritional status of the fetus and the developing brain, thus slowing gestational growth.

Oxygen Deprivation

Known medically as fetal hypoxia, oxygen deprivation is a serious birth complication that has obvious implications in brain development. A study reported in the journal *Pediatrics* found "frequency of oxygen treatment in newborns who later developed autism." Some studies, but not all, have found an increased risk for autism for these conditions related to oxygen deprivation:

- Fetal distress

- Prolonged labor

- Umbilical cord complications

- Caesarean delivery

Birth Abroad

Three studies conducted in Nordic countries found that a mother who gives birth in a foreign country has a "significant" 58 percent increased risk of having a child with autism. We can only speculate why this is. One of the most plausible reasons associates risk with the increased maternal stress of living in a foreign country. Another speculates that women born in another country may not be immunized against the common infections of the country in which she gives birth.

Birth Order

Numerous studies have looked at various factors involving the number of pregnancies and a link with autism. The consensus of opinion has found no link between miscarriage and a risk of autism in a subsequent successful birth. However, there is evidence to support these associations:

Firstborns. Next to the fact that autism strikes boys more than girls by a ratio of 5 to 1 is the fact that autism is most common among

INFANT SURGERY AND AUTISM

Researchers from the Mayo Clinic found a link between surgeries requiring anesthesia before the age of two and healthy brain development. The study compared 350 children who underwent surgery with anesthesia and 700 children who did not undergo any procedure. Among those who underwent two or more surgeries by age two, 37 percent developed a learning disability later in life.

"Our advice to parents considering surgery for a child under age two is to speak with your child's physician," said Randall Flick, MD. "In general, this study should not alter decision making related to surgery in young children. We do not yet have sufficient information to prompt a change in practice and want to avoid problems that may occur as a result of needed procedures. For example, delaying ear surgery for children with repeated ear infections might cause hearing problems that could create learning difficulties later."

firstborns than any other order of birth. The Harvard meta-analysis found that firstborns have a 61 percent increased risk of autism compared to being the third child or later. Risk is second highest for those born fourth or above in order.

I suspect the reason that firstborns are at higher risk has a lot to do with the stress involved in all the unknowns that go with giving birth for the first time. Also many first pregnancies come as a surprise, meaning a woman may not have done anything to prepare herself ahead of time for pregnancy. And most women generally work right up to the time of delivery, which in and of itself is stressful. This generally is not the case for women who already have children.

Close age of siblings. Researchers in California examined the birth records of more than 600,000 second-born children and found that children born less than one year apart had a threefold increased risk of autism than those who were spaced three years apart.

Season of conception. One of the many commonalities among children with autism seems to be season of conception. When researchers at the University of California School of Medicine analyzed data from more than 6.6 million births, they found children conceived in December, February, and March had a 6 percent greater risk of developing autism than children conceived in July. The researchers speculated that seasonal environmental factors—wintertime flu epidemics, for instances—may play a role.

ENVIRONMENTAL POLLUTION

Evidence over the past hundred years has proven that toxic chemicals in the environment can damage the human brain and produce a number of neurodevelopmental disorders, including autism. Yet we keep producing more and more chemicals to the extent that it is just about impossible to test them all for their potential toxicity to life, both human and otherwise.

Today there are more than eighty thousand commercially available chemicals. For the most part, these are all synthetic chemicals that did not exist fifty years ago, when autism was a little-known disorder. These chemicals are found everywhere and have been detected in air, food, and drinking water. It it estimated that more than four billion pounds of pesticides alone—many of them neurotoxic—are used per year in agriculture, a great variety of consumer goods, medications, motor fuels, yard and lawn products, and materials found in and around homes, schools, hospitals, day-care centers, and playgrounds.

Children are particularly at risk of exposure to an estimated three thousand synthetic chemicals that are produced in volumes of more than one million pounds per year. According to the CDC, measurable levels of more than one hundred synthetic chemicals, many of which are neurotoxicants, can be routinely detected in the blood of most Americans of all ages, as well is in umbilical cord blood and breast milk. It is no wonder that both prenatal and early postnatal exposure to a number of chemicals have been linked to neurodevelopmental disabilities.

Over the years, many of these chemicals have been deemed to "be safe at low levels." However, what is safe for an adult is not necessarily safe for an infant or a fetus. Developing brains are much more susceptible to injury from toxic chemicals and other negative environmental factors than an adult brain. We know that many heavy metals can cross the placental barrier. For example, mercury concentration in the umbical cord has been found to be even higher than levels found in the mother's blood.

After birth, a baby becomes even more vulnerable to environmental toxins. The blood-brain barrier, which protects the adult brain from many toxic chemicals, is not completely formed until a baby is six months old. Bottom line: Babies have a relatively higher exposure to chemicals than adults, because of their small size, but less ability than adults to detoxify and excrete them.

Only about half of the high-production chemicals widely used in consumer products and used in the environment have undergone adequate testing to assess their potential toxicity. According to the CDC, less than 20 percent have been tested to specifically ascertain their potential to damage the developing brain and nervous system.

Pesticides and Insecticides

There are almost nine hundred different kinds of pesticides and insecticides used in the United States. A small number of them belong to a class of common insect killers known as organophosphates, which are widely used in agriculture and on the lawns of homes, public buildings, and parks. The two most notorious organochlorine presticides—DDT and PCBs—have been outlawed.

Organophosphates. These kill bugs by disrupting their brains and nervous sytems. However, they can also harm the brains and nervous systems of adults, children, and animals. They are known to be a health risk to farmers who are exposed to them on a routine basis and to children who live on farms. Studies have shown that children from farm communities who were chronically exposed to pesticides had poorer performance on motor speed tests. Now more recent studies indicate they are potentially harmful to children at much lower exposure than originally believed. Studies found that children who were exposed to a mixture of pesticides, including organophosphates in their home environment, displayed short-term memory loss, had impaired hand-to-eye coordination, and were unable to draw simple figures—all signs of a brain imbalance. This was all in the absence of any telltale symptoms of pesticide poisoning. Other children in nearby surroundings who had not been exposed to the pesticides did not exhibit any abnormal symptoms.

Studies have also found that mothers who worked in greenhouses while they were pregnant had children with visuospatial deficits, meaning they have trouble remembering images and feeling the world around them, compared to children who were not exposed to the chemical in the womb.

Organophosphates are hard to avoid. We are exposed to them through foods that have been sprayed with them, drinking water in which they fall, and from breathing the air in which they accumulate. They have an affinity for accumulating in fatty tissue and can be passed to an infant through breast milk. Forty organophosphate pesticides are registered in the United States, with at least seventy-three million pounds used in agricultural and residential settings. These are among those considered the most dangerous:

Chlorpyrifos. This widely used chemical, which is commonly used in agriculture and sprayed to protect the fruit and produce we buy at the supermarket, is under a great deal of public scrutiny by researchers and activist groups. It was developed out of nerve gas. The Environmental Protection Agency (EPA) banned it for home use, but it is still permitted for use on corn, fruit, and other crops, as well as golf courses and treated wood. Common brand names include Dursban and Lorsban.

In 1995, Dow Chemical Company, which makes chlorpyrifos, paid the largest penalty in the history of presticide lawsuits (over $2 million) to the state of New York for advertising that Dursban was "safe." Lawsuits seeking to ban chlorpyrifos for agricultural use are pending.

In my mind, there is no doubt that chlorpyrifos is a dangerous chemical. Consider these findings:

- After the EPA persuaded the major pesticide manufacturers to reduce residential use of chlorpyrifos and diazinon in the United States, there was a decrease in low-birth-weight infants within months after the reduction.

- Children who were prenatally exposed to chlorpyrifos had a smaller-than-normal head size, which is a risk factor for neuro-developmental disorders. Interestingly, however, a smaller head size was seen only in infants whose mothers had a low expression of the enzyme that detoxifies organophospates. Mothers with higher levels of the enzyme were apparently able to clear the chemical out of their bodies, which protected their babies. This is a good example of a (positive) epigenetic effect—the body alters the expression of a gene, which protects the developing brain.

- Children prenatally exposed to chlorpyrifos had significant developmental delays, cognitive deficits, and increased risk for ADHD and pervasive developmental disorder.

- A study of seven-year-olds in California found that prenatal exposure to organophosphates caused a seven-point drop in IQ scores.

Diazinon. This chemical, which is used to kill cockroaches, ants, fleas, and the like, is a known neurotoxin. Although consumer products containing diazinon were banned in 2004, products made prior to the ban are still available. Studies show that even exposure to low doses of this chemical can affect the basic cellular machinery that controls patterns of cell maturation and formation of synapses in the brain.

Malathion. Though long considered to be safe at low levels of exposure, malathion is under scrutiny for being toxic to the developing brain. At one time, the concern over malathion poisoning was focused on ADHD and children who lived in farming communities, but a study conducted by Harvard School of Public Health suggests it is more toxic than previously believed.

The researchers collected data from more than eleven hundred children who lived in areas not exposed to malathion. They found a correlation between children with detectable levels of malathion in their urine and a high risk for developing ADHD. The researchers found the finding "interesting and provocative" because the levels of pesticides found in the children were "very low." The researchers speculated that the malathion in their systems came from food. No surprise, when you consider that a lot of the foods we feed our kids and think of as healthy are tainted with malathion. One federal agency survey found that 28 percent of frozen blueberries, 25 percent of strawberries, and 19 percent of celery were contaminated with malathion.

Polybrominated diphenyl ethers (PBDEs). This flame retardant, which is widely used in building materials, electronics, home furnishings and mattresses, textiles, cars, airplanes, and plastics, has been under scrutiny as a health hazard for more than twenty years. The first human study, conducted by researchers at Columbia University in 2010, found that children who had high concentrations of PBDEs in their umbilical cord blood at birth scored lower on tests of mental and physical development between the ages of one and six than children who were not exposed to the chemical. The results were independent of the mother's own intelligence and other risk factors. Our biggest exposure to PBDEs comes through dust. The study was part of a broader project examining the effects on pregnant women in the aftermath of chemicals released during the destruction of the World Trade Center.

Methyl parathion. This highly controversial pesticide has been banned from indoor use. Studies have found that children exposed to it exhibited short-term memory loss and problems in attention span.

Recommendation: You can't avoid pesticides and insecticides because they are practically ubiquitous in our environment. However, you can reduce your exposure by eating only organic foods and going green in the use of chemicals on your lawn and in your home.

Exposure to Road Pollution

Evidence is mounting that air pollution can affect the brain. Tiny carbon particles as well as other pollutants in tailpipe exhaust from cars, trucks, and buses have long been linked to heart disease, cancer, and respiratory ailments. Now new research is starting to emerge linking it to an increased risk of autism. Numerous public health and laboratory studies suggest that traffic fumes take a measurable toll on mental capacity, intelligence, and emotional stability at every stage of life—and this includes fetuses.

Even though cars and trucks today generate only one-tenth of the

LEAD POISONING AND AUTISM

The toxic effects of lead poisoning and its neurological impact on children is well known, though a case has never been documented associating it with autism.

The detrimental toxic effects lead has on children was first noticed in the 1900s, but it took decades before anyone could prove that the severe headaches, coma, convulsions, and even death in children was a result of lead poisoning. Lead poisoning has been found to have long-lasting neurological consequences. Studies conducted in the 1970s found that children exposed to lead could have decreased intelligence and attention and school problems, even without the presence of any obvious signs or symptoms of lead poisoning. The greater the exposure, the more severe the symptoms. These children are more likely to experience school failure, have dyslexia, and to get arrested.

The interesting thing is that although lead poisoning and exposure were once so pervasive, there didn't seem to be specific cases of autism related to lead. This could be because autism was not readily recognized and diagnosed at that time, or it could be that there needs to be other environmental factors in place to trigger the onset of autism.

Eventually most lead was removed from gasoline, paint, and consumer products. Thanks to bans on lead in consumer products, our exposure today is much lower than in all previous generations. However, it can still be found in municipal water supplies and can leach out of pipes that carry water into homes. Even low levels can create brain problems in children. For this reason, I recommend that pregnant women and women who plan on becoming pregnant have their water tested for lead, or drink bottled spring water.

pollution that they did forty years ago, there are more cars on the road—and more people stuck in traffic—than ever before.

Studies show that breathing street-level fumes for just thirty minutes can intensify electrical activity in brain regions responsible for behavior, personality, and decision-making. Out of the top ten most congested highways, seven are in the Los Angeles area and two are in San Francisco. The one non-California highway is the Van Wyck Expressway in the New York City area.

Studies have found the following:

- California has one of the highest rates of autism and the most congested highways in the nation. Research at the University of Southern California found that children born to mothers living within one thousand feet of a major road or freeway in Los Angeles, San Francisco, and Sacramento were twice as likely to develop autism as children born to mothers living in less polluted cities. These findings were independent of other risk factors, including gender, ethnicity, educational level, maternal age, and tobacco use.

- Premature births, a risk factor for autism as well as other neurological disorders, dropped 11 percent in New Jersey after the state initiated the E-ZPass at highway toll areas to ease up on traffic congestion.

- Children living in areas affected by high levels of exhaust emissions, on average, do not do as well on intelligence tests than children growing up in environments where the air is cleaner, according to separate studies conducted in New York, Boston, Beijing, and Kraków, Poland.

Recommendation: This is a hard one to avoid if you live in one of these areas or any area with lots of traffic, but awareness is important.

Perhaps you might want to consider moving to a more rural area or take frequent trips outside of the city. I recommend getting tests that can measure levels of certain toxic chemicals and stress. You should also consider wearing a face mask when working outdoors, as is commonplace in Asian countries, particularly Japan, to help limit exposure to toxins.

Sensitivity to Heavy Metals

Heavy metals are present in all of us, and have been since the dawn of time. Studies on ancient human bones tell us that cavemen were crawling around with high levels of the same heavy metals we worry about today, such as mercury. Though we read and hear a lot about mercury contamination in fatty fish, there is solid evidence showing that the amount of mercury in fish that swim the world's oceans actually hasn't changed for centuries and, in fact, may be on the decline. There is also convincing evidence that says we are exposed to less

CHEMICAL EXPOSURE AND IQ

Exposure to three environmental chemicals—lead, mercury, and organophosphate pesticides—in the womb and infancy could cost an estimated cumulative loss of 40 million IQ points for America's children, according to an analysis of several studies conducted at Boston Children's Hospital and reported in *Environmental Health Perspectives*.

Lead exposure was by far the most egregious, with a loss of 23 million IQ points, followed by organophosphate pesticides at 17 million points, and methylmercury at 300,000 IQ points. This comes to a 1.6 point drop for every child in the United States under the age of five.

mercury and other heavy metals than many alarmists would like us to think.

There is a lot of talk and research pointing to heavy metals, most specifically mercury, as the smoking gun in the cause of autism. While there is little doubt that certain heavy metals are a risk factor for autism, there is no good evidence that shows they are causing the dramatic rise we are seeing today. Recent studies have found that the levels of mercury in children with autism are actually lower than in healthy children without autism.

The EPA has been stating for decades that the amount of mercury and other heavy metals the typical person is exposed to are not harmful. I tend to agree—*except* for people with an antigen that makes them so hypersensitive to heavy metals that they absorb them almost like a sponge. For them, heavy metal sensitivity is more like an allergy or food sensitivity, as we see with gluten.

We all have high levels of toxins in our bodies. However, it is not the levels of toxins in your body that matters as much as your body's ability to tolerate them. Your ability to tolerate toxins depends on the balance of your immune response, which is controlled by the brain. A brain imbalance can upset the balance of the immune system, which in turn can make you more sensitive to environmental toxins. When we inhale or ingest heavy metals, they move out of the bloodstream and settle deep into our tissues. People with high levels of heavy metals can go through a therapy called chelation—and many people do— that leaches out and binds with the metals and takes them out of the body through the elimination canal. However, in people who are sensitive to heavy metals, this method can actually backfire and make the problem worse.

We can develop an allergy or hypersensitivity to just about anything, and I believe heavy metals is one of them. I also believe that this is the way they become a risk factor in autism. It's not toxicity per se that is the troublemaker, but an individual's own immune response.

A compromised immune system is a common thread in children with autism, and one they share with one or both parents. This, I believe, is how heavy metals become a risk factor for autism.

Researchers in San Francisco studied more than 250 children with and without autism in the Bay Area who live in communities known to have high emissions of heavy metals. They targeted nineteen chemicals suspected of having a link to autism and came up with five suspects with a "potential association" to an increased risk. They are mercury, cadmium, nickel, trichloroethylene, and vinyl chloride. Here's what you should know about them:

Mercury. High levels of mercury can be dangerous. Several years ago two thousand people in Japan came down with mercury poisoning as a result of exposure to industrial waste containing mercury. There was a similar occurrence in the Philippines, where people ate fish from a body of water contaminated with industrial waste. These are examples of industrial accidents rather than everyday exposure. Nevertheless, mercury poisoning has been linked to brain damage, poor memory, learning problems, and delays in speech and reading ability in children. When researchers in Texas looked at the rate of autism in connection to mercury, they found the rate of autism was 2.6 percent greater for every thousand pounds of mercury released in the vicinity of the geographic center of 1,040 school districts. It was 3.7 percent higher in children living near power plants. In addition to exposure from industrial emissions and eating fish, mercury is found in scientific apparatus, such as thermometers and barometers, and compact fluorescent lights.

Cadmium. This metal has been widely used to make such things as plastics, steel plating, batteries, and solar panels. It is also used in nuclear fission. Improved technologies have reduced the demand for cadmium, so it is not as commonly used as it once was.

Trichloroethylene. This chlorinated hydrocarbon is an industrial solvent that can seep into groundwater. It is used in solvents, such as those used in dry cleaning, refrigerants, and to extract oils from plants.

Vinyl chloride. This is a known carcinogen that is not in wide use today. It was once used in hair sprays and other aerosol sprays. It is considered an intermediate compound, meaning it is not used in the final production of products. Our exposure to it comes from the emissions from plants that manufacture it.

Recommendation: You can't completely eliminate your risk, but you can lower it by being aware of where these metals exist in your environment. You can also have your water tested for heavy metals. The level of the detoxifying chemical glutathione in your body is the main predictor of your immune chemical tolerance. Making sure this chemical is high in your body is your best protection. Supplements called glutathione recyclers can help keep your level of this chemical high. You can find out about how to test for it in Chapter 10.

ELEVATED HORMONE LEVELS

Sex hormones are very active in the womb. During fetal development, hormones play a key role in determining if your child will be male or female. They are also very important in the development of the brain. Studies show there is a correlation between abnormal levels of these hormones in both the womb and after birth and an increased risk of autism. Studies also show that an increased stress response is associated with abnormal hormone levels in pregnant women—yet another way that stress contributes to the risk of autism.

Testosterone and Estrogen

Autism is a male-dominated condition, which is why there is so much attention on the male hormone testosterone. Several years ago, research led by the renowned British autism researcher Simon Baron-Cohen found that high testosterone levels in amniotic fluid can be found in mothers of children who end up with autism. New research now offers clues as to why high testosterone levels can raise the risk of autism. It appears that testosterone and estrogen have opposite effects on a very important gene nicknamed RORA, which plays a key role in regulating other genes involved in brain development. They found that testosterone suppresses the RORA gene, while estrogen turns it on. When a cell has high levels of testosterone, RORA levels run low, which impacts every gene that RORA is supposed to turn on. It's important to note that the study did not show that low levels of RORA cause autism, only that they are associated with the condition. However, other research suggests that a RORA deficiency might explain many of the effects seen in autism. For example, studies have found that:

- The RORA gene protects neurons against the negative effects of inflammation and a heightened stress response, both of which are known to be elevated in autism.

- Brain tissue in people with autism contains lower levels of RORA than in people without the condition.

- RORA may help maintain the body's daily circadian rhythm. People with autism frequently experience sleep disturbances, which can be caused by an imbalance in the hormones that regulate the body's natural rhythms.

- Mice that have been genetically engineered to lack the RORA gene engage in a number of autistic-like behaviors.

Unlike testosterone, estrogen raises RORA levels. This may help explain why females are protected against autism. Even if RORA levels were otherwise low, estrogen can "pick up some of the slack." Researchers are not saying that RORA is the only gene involved in autism, but it's likely one of the critical ones. It's important to note that males, but not females, are sensitive to the effects of hormones in the womb. We all start out as females at conception. The female form changes to male as a result of the secretion of high levels of testosterone and lower levels of estrogen. In the absence of any hormones, a female form still will take shape. This also means that the brain of the male fetus is much more vulnerable to any environmental influence that may interfere with gene expression and the proper secretion of hormones.

Oxytocin

Oxytocin is the hormone of emotional bonding. It is called the love hormone because it is secreted to produce an orgasm when making love, it helps produce milk for a suckling infant, and it triggers the biological process that starts labor. Emotional bonding is counterintuitive in autism, and scientists speculate low levels of oxytocin might play a role. In fact, some research is looking at the possibility of using oxytocin as therapy to treat the social deficits of autism.

There is a close relationship between oxytocin and the stress response. Studies show that oxytocin helps reduce cortisol levels and thus dampens the stress response. Likewise, it is also believed that a high stress response decreases the hormone of human bonding, which is why some scientists speculate it may play a role in autism. Though there are no studies that show this, scientists base their hunch on the unfortunate results of using a synthetic version of oxytocin, a drug called Pitocin, that came on the market in 1955 to help induce or speed up labor. Pitocin was used because mothers had inadequate levels of oxytocin to aid labor. Some reports show that as many as 60 percent of children with autism also were exposed to Pitocin. However, I believe there was more in play here than just exposure to Pitocin. The need to induce labor implies a difficult birth, which would elevate the mother's stress response and inhibit the production of oxytocin. This is a relationship that needs more investigation.

Recommendation: Hormone testing is a rather simple procedure, which you can read about in Chapter 10. However, I believe the key to healthy regulation of hormones in the womb is by managing your stress response. An increased stress response is associated with abnormal hormone levels in pregnant women.

INFLAMMATION AND A MOTHER'S IMMUNITY

Inflammation makes for a hostile fetal environment. At one time the relationship between autism and an aberrant immune system was a radical notion, but not anymore. Studies are finding that a subset of children with autism and their mothers share high levels of inflammation-producing immune antibodies that researchers believe cross the placenta from mother to fetus and affect the fetal brain. More proof of this comes from studies that show certain patterns of these antibodies are found only in mothers of children with autism, but not in mothers of children without autism.

An inflammatory immune response in mother and child is one of the hottest and most pursued areas of autism research. Maternal antibodies called immunoglobulin G (IgG) are naturally transferred across the placenta to the fetus throughout pregnancy because they will serve an important protective role until the child's immune system matures at around six months. However, we are now seeing that other antibodies that can be harmful to brain development can have equal access. This appears to happen in mothers who have chronic inflammation caused by an overactive immune system. However, most recent research also suggests a genetic link.

A study involving three hundred pregnant women conducted at

the University of California's MIND Institute found that women carrying a particular gene called MET are more likely to produce antibodies that may injure the brain of a developing fetus and increase the risk of developing autism. These brain-affecting antibodies were found in mothers of some of the children who were later diagnosed with autistim, but they were not present in mothers of normally developing children.

"Our study gives strong support for the idea that, in at least some cases, autism results from maternal immunity gone overboard," says Judy Van de Water, one of the lead researchers. "Our study has found that a kind of safety switch that regulates the immune system and prevents it from targeting the brain of the developing fetus is defective in some mothers of children who later develop autism." An overactive immune system is often associated with a right-hemisphere defect. It is possible that both may occur together, and that the combination would increase the risk of autism.

Here are some of the other findings:

- A study reported in *Annals of Neurology* was one of the first human studies to show that maternal antibodies could potentially cause autism. The researchers measured increased levels of antibodies in the mothers of three children. One child had autism, a second child had a language disorder, and a third child had no neurological problem. They took the antibodies from each of the mothers and injected them into three sets of pregnant rodents. The offspring of the mice injected with the antibodies of the mother with autism displayed autistic-like symptoms compared to the others.

- Other studies have found patterns of antibody reactivity in mothers of children with autism from two to eighteen years after giving birth.

- Autopsies of adults with autism have found evidence of lifelong chronic inflammation in the brain. They showed excessive growth of white matter in the immune cells of the brain known as microglial cells. Research at Johns Hopkins University School of Medicine suggests that low-grade inflammation in pregnant women can pass through the placenta and blood-brain barrier and cause aberrant growth in the brain of a fetus that can lead to autism.

- Research at the New York Institute for Basic Research in Developmental Disabilities concluded that inflammation and a type of cell death called apoptosis "may play a significant role" in the development of autism.

- One study discovered that autoantibodies, which are proteins in the immune system directed at itself, target brains in both children with autism and their mothers. "This finding has great potential as a biomarker for disease risk and may provide an avenue for future therapeutics and prevention," says Dr. Datis Kharrazian, a world-renowned expert on autoimmune disease and the immune system.

Other findings have identified widespread changes in the immune systems of children with autism, including signs of active, ongoing inflammation, as well as alterations in genetic pathways associated with immune signaling and immune function. Also, many genetic studies have indicated links between autism and genes that are important to the nervous system and the immune system.

It's clear why this happens. Our immune system contains important protective agents that come to our defense when foreign invaders known as antigens set out to do us harm. However, everything in life is about balance, and our immune system is no exception. A healthy immune system is dynamic, able to respond swiftly and appropriately to every threat we are faced with by switching back and forth as needed between two pathways, one that is pro-inflammatory, and one that is anti-inflammatory. In people with autism, this does not appear to be the case. Instead, we see an overactive pro-inflammatory response. Studies investigating the immune systems of children with autism have found high levels of inflammatory cytokines. Studies have also found the same thing in their mothers.

However, I do not see this as a problem with the immune system per se, but with how the brain regulates it. The brain has an active role in controlling immune response, a little-known fact that nonetheless has been well documented in the scientific literature over the past thirty years. The left brain activates the immune response and the right hemisphere inhibits it. Because the left brain is too strong and the right brain is too weak in autism, we see the same thing in the immune system. I see this as a very important key in autism.

Inflammation is caused by an immune system in high gear. It is so on the ready to attack that it has a hard time distinguishing between a harmful invader and its own cells. It's the immune system's version of friendly fire. This is what puts mothers at risk for infection during pregnancy and why we can develop a sensitivity to heavy metals that we can pass on to our children. It explains why having an autoimmune disease may raise the risk of having a child with autism. I believe and the research proves that, in most cases, this inflammation is the result of this imbalance in the brain and not the cause of it. For example, when Harvard's Dr. Herbert found an unevenness of function characteristic of the autistic brain, she also found an overgrowth of cells associated with inflammation everywhere in the brain. A brain imbalance causes an imbalance in the immune response, which increases inflammation everywhere in the autistic body and brain. Inflammation is associated with a host of other immune problems we see in people with autism, including food sensitivities, and a persistent and problematic stomach ailment known as leaky gut syndrome.

FOOD ADDITIVES: IS THERE A LINK?

Contrary to what some people might believe, there is no good evidence to support a link between sugar and neurobehavioral disorders, but there is for the food coloring that goes into making candy.

There is wide concern that a variety of chemicals used in food manufacturing are causing some of the behavior problems we are seeing today in kids. Most notable among them is food coloring. While there has not been much written about autism and food coloring, there has been significant research relating eating foods containing food coloring to the other well-known right-brain deficiency, ADHD.

The initial concern over food coloring and behavior was spurred in 1973 by pediatrician Ben Feingold, MD, who told a meeting of the American Medical Association that food additives were responsible for 40 to 50 percent of the hyperactivity he observed in children in his practice. Subsequent studies examining Feingold's additive-free diet have been mixed, so support of this approach is theoretical at best. However, there have been plenty of other studies on food additives.

One of the most well-known and well-conducted studies to date, published in the British medical journal *The Lancet*, looked for a link between the food additives and behavior among 155 three-year-olds and 144 eight- and nine-year-olds. After an initial period in which the children were put on an additive-free diet, they were divided up and given a daily juice drink containing sodium benzoate, a common food additive, and a dye, or a placebo. The researchers found a significant increase in hyperactivity in the kids drinking the juice containing additives. However, what isn't known is if the behavior was caused by the sodium benzoate, the additives, or both.

Recommendation: I think there is enough good research linking food additives to a right-brain deficiency that pregnant women and women who are considering becoming pregnant should exercise caution and avoid them as much as possible.

Recommendation: Testing for inflammation and antigens is fairly simple. I'll tell you how to go about it in Chapter 10.

Food Sensitivities

We're all familiar with food allergies, the type in which eating a peanut or other food can trigger an immediate and sometimes life-threatening reaction. Food sensitivities are quite different because they are not easy to find. They don't create a life-threatening situation, but they are harmful to us in a more subtle way—they create chronic inflammation.

Food allergies and food sensitivities develop along different pathways. Food allergies develop along the anti-inflammatory pathway. The body recognizes a particular food as an enemy invader and responds by sending out histamine to quash the invader. A food sensitivity is different because it triggers no immediate action. Rather, an offending food acts more like an irritant and rouses IgG antibodies along the pro-inflammatory pathway. The reaction is subtle and can take anywhere from a few hours to several days to respond in the body. The symptoms are also subtle and can be nothing more than bloating, gas, or feeling sluggish or moody. The menacing effect is what we can't feel—inflammation, the kind that breeds bad health. These IgG antibodies are the same ones that have been shown to cross the placenta and enter the fetal brain.

You can become sensitive to just about any food, but the most common sensitivities we are seeing today are to gluten, a protein-like substance found in wheat and wheat products, and casein, the protein in milk.

Recommendation: Testing for a food sensitivity involves a bit of detective work, but I will show you how to do it—and help you determine whether you should—in Chapter 10.

Leaky Gut Syndrome

The right brain is primarily in charge of digestion. So if you have a brain imbalance where the right side is weak, you will also have digestive ills. Digestive complaints are chronic problems in children with autism, and I am finding more and more that their parents have similar complaints. Often, when parents describe their child's digestive problems they add, "He's a lot like me, but worse."

One manifestation we commonly see is an inflammatory condition called leaky gut syndrome. A healthy stomach is lined with closely linked cells that form a tight barrier against only the smallest of molecules. This protects the stomach against foreign invaders, such as bacteria, viruses, fungi, and other parasites.

Decreased brain activity reduces blood flow to the stomach lining as well as decreasing the production of hydrochloric acid in the stomach, which aids digestion. Less hydrochloric acid means that food, especially protein, remains partially undigested and harder to absorb. Reduced blood flow to the gut interferes with absorption of all nutrients and can lead to a thinning and eventual breakdown of the gut lining. This is leaky gut syndrome.

The lack of acid in the stomach promotes the overgrowth of "bad" bacteria, yeast, and fungus, all of which can produce toxins that can now easily leak through the intestinal lining. Increased tone in the stomach lining can reduce the production of pancreatic enzymes that are needed to detoxify chemicals. This leads to a buildup of toxins. Folate and vitamin B_{12} are essential to the production of glutathione, one of the body's main detoxifying chemicals. Poor glutathione production interferes with the body's ability to produce digestive enzymes. Lack of digestive enzymes also reduces the ability to break down protein to amino acids, which are needed to make neurotransmitters. Protein that can't be fully digested will affect brain chemistry. This is why some children with autism will develop an addiction to a particular food—such as, *I want pasta for breakfast, lunch, and dinner*. And it is usually a food they are sensitive to. As you can see, it is a vicious cycle. Leaky gut is a classic example of the downstream effect that we see as many of the symptoms of autism.

Recommendation: Preliminary research indicates that antibody abnormalities in the mother could account for possibly 12 percent of cases of autism. I considered this strong evidence, and for this reason I suggest that all men and women contemplating pregnancy should have their immune response examined. I show you what it takes in Chapter 9.

THE PRICE OF A MODERN LIFESTYLE

I have just laid out nearly three dozen risk factors that can affect the body in a way that could increase the risk of having a child with autism. The logical question to ask is *why*. Why is all this a problem now? Why did these problems not exist thirty or fifty years ago? What is triggering our body to respond in ways that compromise our immune system, increase inflammation, stoke the stress response, and create all the biological mechanisms that have been found to play a role in autism? The answer is our lifestyle.

As I noted at the start of this book, studies show that more than half of the dramatic rise in autism cannot be explained by genetics alone. I believe it has a great deal to do with our modern society and our lifestyle as a whole, which has changed more dramatically over the past twenty years than any time in history. During this span of time our lifestyle has been riding a wave of change side by side with the rising prevalence of autism. I do not believe it is coincidental. There is plenty of evidence pointing to a link between the two. I believe that many of the lifestyle practices commonplace today are fueling a genetic predisposition and creating the eipgenetic conditions that are making brain-building genes misbehave. Here is what the evidence shows:

The Diet of the Mother- and Father-to-Be

It doesn't take a stretch of the imagination to realize the importance of diet in a healthy pregnancy. However, most people don't realize how far-reaching the effects of a bad diet can be. Diet is one of the major ways in which epigenetics can change the course of gene expression. Several studies have found that the wrong kind of diet, both in men and in women, can pass on methylated genes to future generations, even when the same eating style is not repeated in subsequent generations. In one study, researchers, using well-kept harvest records from a small Swedish community, found that exposure to either feast or famine in a group of boys during the slow-growth period prior to puberty affected the longevity of the grandchildren they had decades later. Shortage of food during the grandfathers' boyhood was associated with extended lifespan in their grandchildren, and food abundance was associated with a greatly shortened lifespan due to chronic disease in their grandchildren. When the researchers examined the difference in the grandchildren who lived longer versus those who died at a younger age, the difference was an astounding thirty-two years.

"Epigenetic transmission from just father to child would be sufficient to set up a cascade of metabolic responses down the generations," wrote geneticist Marcus E. Pembrey, MD, in a commentary on the study published in the *European Journal of Human Genetics*. The article was appropriately titled "Time to Take Epigenetic Inheritance Seriously."

In another study using test animals, researchers from Duke University found that the diet of a mother can have an epigenetic effect on the health and behavior of their offspring similar to what we see in autism. The diet-induced changes continued to affect future generations, even though the same diet was not replicated beyond the first-generation mothers.

A long-term study conducted in the Netherlands that started at the end of World War II found that children who were exposed to famine in the womb during the second and third trimesters were abnormally small at birth but ended up with lifelong obesity and other health problems. They had roughly twice the obesity rate of the general population. Fetuses exposed to the famine starting in the first trimester were larger than average, suggesting that the mothers' bodies initially tried to compensate for lack of food in some way, perhaps increasing the size of the placenta.

Recommendation: Studies such as these are a warning that good dietary habits are something that should be started by prospective moms and dads long before conception. You can find out more about foods you should eat and supplements you can take in Chapters 9 and 10.

A Sedentary Lifestyle

The positive effects of an active lifestyle cannot be overstated. Most doctors would agree that if a choice had to be made between recommending exercise or a healthy diet, exercise would come out the winner. I am not going to go into all the negative ramifications of a sedentary lifestyle here, as you probably are aware of them already. What you may not realize is the importance of movement to the developing brain. Movement, particularly using the large muscles, is a major stimulant for proper brain development. Movement makes the brain grow. It's why kids need to be outside climbing trees and kicking balls instead of playing computer games and texting friends. The same goes for keeping the adult brain healthy.

Recommendation: It is not as easy as just saying "start moving." I explain why and how to work a productive exercise routine into your life in Chapter 10.

Vitamin D Deficiency

With the accent on too much of just about everything in our modern society, it seems surprising that an old nutritional *deficiency* is resurfacing, and at an alarming rate: vitamin D deficiency. The main reasons for this is lack of sun exposure, overzealous use of sunscreen, and less time spent in the great outdoors.

"The prevalence of autism over the last twenty years corresponds with increasing medical advice to avoid the sun, advice that has probably lowered vitamin D levels and would theoretically greatly lower activated vitamin D levels in the developing brain," says Dr. Datis Kharrazian, one of the world's leading experts on autoimmunity and immune dysfunction.

Recent research has found that the benefits of vitamin D have been greatly underestimated, and lack of adequate vitamin D is associated with a growing number of health problems, including autism. Research implicates a vitamin D deficiency in autism in two ways: by compromising the mother's immune system and by affecting fetal brain development.

Vitamin D deficiency in parents prior to conception was identified as a key risk factor for autism in a major review conducted by researchers from Harvard Medical School, Boston Children's Hospital, and McLean Hospital in Belmont, Massachusetts. The researchers even suggested that supplementing the diet with vitamin D prior to conception may someday play a leading role in reducing the risk of having a child with autism.

Animal studies have repeatedly shown that severe vitamin D deficiency during pregnancy results in offspring with brain abnormalities similar to those found in autistic children.

Even if you get adequate sun exposure, there are factors that could put you at risk for vitamin D deficiency, including leaking gut or other inflammatory gut conditions and a high stress response. Dark-skinned individuals and people who consistently use sunblock are also at risk. We also lose some of our ability to absorb the vitamin as we age.

Recommendation: The only way to get vitamin D naturally is from the sun. Vitamin D is fortified in many foods, especially milk. Most doctors these days recommend taking vitamin D supplements. I suggest that both men and women who are planning to conceive take from 1,000 IUs to 4,000 IUs a day. Some experts recommend as high as 8,000 IUs a day. Women and prospective fathers should have their blood tested for vitamin D level prior to pregnancy. I believe the level for healthy pregnancy should be fifty or above.

Where You Live

Studies have found a higher rate of autism for parents who live in these environmental settings:

- Urban rather than rural

- Geographic regions at high latitude

- Areas of high precipitation

- Close proximity to power plants

Recommendation: You can't necessarily change where you live, but you can take other measures that can help you reduce your risk. That is what the rest of this book is all about.

Too Much TV Time

One of the biggest deterrents that gets in the way of children being active outdoors is the television. One study conducted by the National Bureau of Economic Research found that "early childhood TV watching is an important trigger for the onset of autism." When the researchers

AUTISM, STATE BY STATE

Here's how the fifty states rank in the prevalence of autism. The list, complied by FightingAutism.com, is based on public school records of eight-year-olds. Forty-seven states have rates lower than the national average.

1. Minnesota: 1 in 67
2. Oregon: 1 in 77
3. Maine: 1 in 80
4. Connecticut: 1 in 99
5. Rhode Island: 1 in 104
6. Pennsylvania: 1 in 105
7. Indiana: 1 in 106
8. Massachusetts: 1 in 107
9. New Jersey: 1 in 114
10. Wisconsin: 1 in 120
11. Nevada: 1 in 124
12. Virginia: 1 in 127
13. Maryland: 1 in 127
14. California and Michigan: 1 in 128
15. New Hampshire: 1 in 138
16. Missouri and North Carolina: 1 in 140
17. Georgia: 1 in 145
18. Wyoming: 1 in 149
19. New York: 1 in 151
20. Arizona: 1 in 153
21. Florida: 1 in 155
22. Ohio and Illinois: 1 in 158
23. Washington: 1 in 159
24. Delaware: 1 in 160
25. Texas: 1 in 163
26. Nebraska: 1 in 168
27. Utah: 1 in 175
28. Washington, DC: 1 in 178
29. Kentucky: 1 in 181
30. Idaho: 1 in 185
31. West Virginia: 1 in 193
32. South Dakota: 1 in 199
33. Tennessee: 1 in 205
34. Hawaii: 1 in 206
35. Alaska: 1 in 212
36. South Carolina: 1 in 217
37. Kansas and North Dakota: 1 in 219
38. Arkansas: 1 in 224
39. Alabama: 1 in 227
40. Montana: 1 in 233
41. Colorado: 1 in 273
42. New Mexico: 1 in 275
43. Louisiana: 1 in 295
44. Oklahoma: 1 in 309
45. Mississippi: 1 in 317
46. Iowa: 1 in 718
47. Vermont: figures not available

examined data in California and Pennsylvania between 1972 and 1989, they found a direct correlation between the increase in children with autism and the rising popularity of cable television. That may not surprise a lot of people. However, their data are also consistent with autism risk being associated with too much time spent inside watching television by *parents*. Parents are role models for their children. If you spend too much time in front of the television, chances are your children will too.

Recommendation: Limit your television viewing and your children's to no more than an hour or two a day.

Lifestyle Diseases

More and more research indicates that chronic inflammation and a high stress response generally go hand in hand. If you have one, chances are good that you have the other. More and more research shows that they are key instigators in many of the chronic health problems now labeled "lifestyle diseases." However, the symptoms are not always readily apparent. If you have any of the following conditions, it is likely that you also have chronic inflammation and a high stress response:

- High blood pressure
- A fast resting heart rate, generally considered to be higher than ninety beats per minute
- Unusual sweating
- Prediabetes or diabetes
- Depression and frequent anxiety
- Overweight or obesity

You can find out how to measure for these problems and what to do to get them under control in Chapters 8 and 9.

RISK FACTORS AND AUTISM

To summarize, there is documented evidence showing a link between these environmental factors and autism. These risks are for the prospective mother unless otherwise indicated.

1. One or both parents with strong left-brain skills and high intelligence.
2. An undetected brain imbalance in one or both parents.
3. A mother older than thirty-five and/or a father forty or older. The risk is higher for an older dad than an older mom.
4. A mother-to-be who is overweight or obese at the time of conception and/or is considered to be obese at the time of giving birth.
5. Overweight and sedentary parents.
6. Parents who display mild autistic-like personality traits.
7. Parents with a psychiatric history, especially the father.
8. Maternal stress and a chronically high stress response in the mother or possibly both parents.
9. Chronic inflammation in the mothers or possibly both parents.
10. An overactive immune system in both parents.
11. Elevated hormone levels, especially testosterone, during pregnancy.
12. Vitamin D deficiency.
13. Type 1 (insulin-dependent) diabetes in mothers or fathers.
14. Gestational or type 2 diabetes at the time of pregnancy.
15. Gluten intolerance in the mother or possibly the father.
16. Leaky gut syndrome.
17. Rheumatoid arthritis.
18. Taking these three drugs in early pregnancy and possibly before: thalidomide, misoprostol, valproic acid.
19. Taking antidepressant medications before and during early pregnancy.
20. Rubella (German measles) infection in early pregnancy.

21. Getting the flu, another viral infection, or a bacterial infection during pregnancy.
22. Getting a fever during pregnancy.
23. Folate deficiency or a compromised ability to absorb folate.
24. Hyperemesis gravidarum, an extreme form of morning sickness.
25. Bleeding during pregnancy.
26. Traumatic birth resulting in:

- Premature birth
- Low birth weight (less than five pounds, eight ounces)
- Small for gestational age
- Low APGAR score
- Oxygen deprivation
- Breech or caesarean birth
- Fetal distress
- Prolonged labor
- Umbilical cord complications

27. Giving birth abroad.
28. Being the firstborn.
29. Close age of siblings.
30. Winter birth.
31. Exposure to pesticides and insecticides.
32. Sensitivity to heavy metals.
33. Exposure to road pollution.
34. High blood pressure.

PART 2

FINDING
SOLUTIONS

CHAPTER FIVE

What's Your Cognitive Style?

The notion that we are either right-brained or left-brained came about in the 1960s, when scientists for the first time were able to run separate tests on each side of the brain. What they found was a revelation: There is a big difference in what the two sides do. When the research was reported in the popular press, it was oversimplified and sensationalized, sparking a fascination about being either left-brained—the analytical type—or right-brained—the creative type.

Virtually everyone has personality traits that make them either right-brained or left-brained—what I collectively call your cognitive style. However, this doesn't mean you only use one side of your brain when thinking, performing a task, making a decision, or learning. This is impossible because it takes both sides of the brain working together to effectively and efficiently do everything, from reading a book to playing an instrument to walking down the street. However, virtually everyone tends to tip the scale toward either right-brain or left-brain skills, hence the moniker right-brained or left-brained.

For example, both sides of the brain are involved in speech and language, but the left side is responsible for verbal language—spelling, vocabulary, reading, the literal meaning of words—and the right side is responsible for nonverbal communication—eye contact, gestures, posture, tone of voice, the timing and intensity of speech, and reading comprehension.

Both sides of the brain are creative, but in different ways. For example, the left side guides detailed imaging, which makes painting a still life or a portrait a left-brain skill. The left-brained artist looks at reality and tries to copy it in as detailed a way as possible. Music for the left brain is similar. Most great classical musicians are extremely left-brained. They are able to finely manipulate their fingers (piano or guitar) or mouth (the sax or tuba).

Right-brained artists are better at abstracts and will create something out of their own mind that is unusual and does not actually exist in reality. Right-brained musicians are generally the songwriters. Their right brain guides them to compose lyrics or take old music and arrange it in a new and unusual way.

Understanding the functional responsibility of the two sides of the brain can help you figure out which side of your brain is stronger. With rare exceptions, we are all better at some things than others due to slight imbalances in the brain that label us either right-brained or left-brained. We get this way based partly on our genetic makeup and partly on our experiences. A brain in perfect balance and harmony is truly remarkable and rare. These are our geniuses, the Leonardo da Vincis and Steve Jobses of the world. The other extreme, where one side is very strong and the other side is very weak, are individuals with autism.

SIZING UP YOUR COGNITIVE STYLE

It doesn't take autistic parents to pass on genes that can lead to autism. As you learned in Chapter 2, the right combination of normal genes turned in the off position can produce a child who will develop autism. However, it isn't totally haphazard, as there are some telling clues as to whether or not you may possess the genes that make you an autism phenotype, and therefore increases your risk of having a child with autism. You should be able to figure it out by assessing your own cognitive style. That is, are you more left-brained than right-brained? Do you have autistic traits—weak right-brain abilities and/or possibly strong left-brain abilities—but on the mild or subtle side? As you learned in Chapter 5, people with a left-brain cognitive style are more at risk of having a child with autism than people with a right-brain cognitive style. The more left-brained you are, the more likely you are to possess the genes that can interact with the environment to produce autism.

It is possible for anyone to have personality traits that mildly reflect the symptoms of autism. If you do, it means you can pass these same traits on to your children, which possibly could increase their risk of autism. When two people with mild autistic traits conceive, the risk that their child will develop autism gets even higher. Exposure to certain environmental factors that are known or suspected risk factors for autism raises the risk even more, and so on down the line.

Your phenotype, as I explained in Chapter 2, is the result of the way your own genes were expressed and interacted with the environment to give you your individuality—your physique, your intellect, your personality, your mannerisms, as well as other traits, and the strengths or weaknesses that are a part of them. Most children with autism have parents, siblings, and even other relatives who have autistic traits themselves, but to a much lesser degree. These traits fall into three categories:

- **Socialization skills**—Autistic people are socially withdrawn, some to the point where they have virtually no social skills. An autistic phenotype, or someone with a left-brain cognitive style, is more of an introvert who rarely goes to parties and does not like to throw them. They are the proverbial wallflowers of the world.

- **Communication style**—Autistic people are not expressive at all or have difficulty with some aspect of communication. About 25 percent are nonverbal. The rest verbalize in somewhat of an odd way. They speak with in a flat, robotic manner lacking any expression. Virtually everyone with autism struggles with non-verbal communication—they can't read people and don't under-stand emotion. They don't get facial expressions, they lack eye contact, and they can't interpret body posture or gestures. They tend to be very literal. You might never call an autistic pheno-type a great communicator, unless you get on a subject of par-ticular interest to him or her. Then he might astound you with what he knows and how well he says it.

- **Stereotypic behavior**—Repetitive or obsessive behavior is a common trait in autism. This could be repetitive movements (known as stereotypies), such as hand flapping or rocking, or compulsive actions. People with autism may obsess on a par-ticular topic, or repeat the same phrases over and over again—so-called scripting. They love familiarity or sameness and hate to transition to new things or topics. Tics, even Tourette's-like symptoms, are not unusual. Bill Gates is said to be someone who rocks frequently. It's been written that anyone who spends a lot of time with him starts to rock back and forth themselves. I would say that Bill Gates has autistic traits and a very strong left-brain cognitive style.

Figuring out if you have an autistic phenotype is somewhat subjective. No line can be drawn for what crosses from normal to mild autistic traits. Your cognitive style and whether or not you possess mild autistic traits will be subjective, based on your own assessment and the judgment of those who know you best. But here's the big clue to get you started on your self-assessment: The combination of strengths and weaknesses that we typically see in children with autism, especially those with high-functioning autism such as Asperger's, perfectly reflects a weakness or immaturity in the right hemisphere of the brain and an unusually strong left hemisphere.

It should be obvious from everything you've read so far that the autistic phenotype is someone with a dominant left-brain cognitive style. We all have a cognitive style, just like we are either right-handed or left-handed. We're either more right-brained or more left-brained. No one is all of one and none of the other. We fall somewhere on a continuum of what is termed mixed laterality, or our ability to favor right or left. Handedness is the most obvious example. Most people are right-handed and favor it to do almost any task. However, the left hand is still quite functional and can, say, catch a fly ball headed its way. For some people, attempting to use the opposite hand to do anything turns into a disaster. This represents the two extremes on the spectrum. However, most of us fall somewhere in between. Cognitive style—right brain versus left brain—works the same way. You use both sides of your body, but you depend on one side more than the other. Children with strong left-brain dominance are on the autism spectrum. The more severe the left-brain dominance and the weaker the right-brain development, the greater the severity of autism symptoms. Just as two left-handed parents are more likely to have left-handed children, parents who have a left-brain cognitive style are likely to have children with even stronger left-hemisphere skills. Combine this with other environmental factors that are believed to cause a developmental delay on the right side of the brain while in the womb or within the first three years of life—the years of strong right-brain development—and the risk that their child or children will get autism grows. I believe that parents who are most susceptible to meet the criteria for having a child with autism are those who are highly gifted in left-brain skills—what is commonly referred to as being left-brained.

Knowing your cognitive type is important knowledge for both you and your spouse or partner to have if you are planning on having children. If you already have children, you should still assess yourselves and them as well.

An important piece of the puzzle is knowing the typical symptoms of autism, so you can recognize the traits in yourself. Review the following list of autistic traits and think about them as they might apply to you now or when you were a child. Of course, think in terms of them being mild or so subtle that they are barely noticeable or may only be known to yourself or the person you are the closest to. I've categorized the traits as they pertain to autism: weak right-brain skills or strong left-brain skills. This is not a test. That comes later. Just file the information away and continue on to take the quiz that will help you define your cognitive style.

DOES THE SHOE FIT?

Might you have some autistic traits, or did you display them as a child? Below you will find the more common traits we look for when diagnosing autism. The actual diagnostic criteria used to diagnose autism are quite extensive, so this is an abbreviated version. I've categorized these traits according to the hemispheric problem that produces the symptoms, which diagnosticians generally do not consider. At our Brain Balance Achievement Centers we believe that this approach is essential for finding an effective treatment strategy.

These Traits Indicate a Right-Brain Deficiency, Which Is Caused by the Hemisphere Growing Too Slowly

- Little or no eye contact.

- Severe hearing difficulty to the point of appearing deaf, even though hearing tests fall within the normal range. For example, a child who does not respond to his or her name.

- Delayed speech or a complete absence of speech.

- Delayed walking.

- Walks and runs with an awkward gait.

- Prefers isolation and appears to be in a world of his or her own.

- Rarely smiles and wears a blank expression.

- Resists cuddling, affection, and close contact.

- Is an extremely picky eater.

- Lacks or has a reduced sense of smell.

- Has a poor detoxification system.

- Has digestion problems and inflammation of the gut, known as leaky gut syndrome. This can actually fit both a right-brain deficit and a left-brain strength, but is more common as a right-sided deficit.

These Traits Indicate a Left Brain That Is Too Strong as a Result of Growing Too Fast

- Repetitively flaps hands, waves arms, rocks back and forth, or spins.

- Repeats verbatim what others say.

- Engages in repetitive behavior and/or routines.

- Plays with toys inappropriately, such as holding a car upside down and spinning the wheels.

- Displays a preference for order, such as lining up toys.

- Shows no separation anxiety when parents leave.

- Is hypersensitive to sensory stimulation, such as sound, light, touch, etc.

- Pounds head against the wall or another object.

- Does not play pretend or engage in imaginative play.

- Has low muscle tone and delays in large muscle skills.

- Appears fearless and does not worry about being around strangers.

- Has unusual and/or limited interests.

- Is impatient and can't sit still.

- Does not like change and will resist it.

- Started reading and writing at an unusually young age, under two years.

- Is visually hyperfocused on details.

- Has an extremely good memory for details, numbers, dates, etc.

- Has a superior ability to draw in fine detail.

- Has perfect pitch—a natural ability for music.

- Has chronic inflammation in both the body and brain.

- Has autoimmune problems, such as eczema, asthma, or allergies.

- Has food sensitivities (which are different from food allergies, and more difficult to detect).

WHAT DOES BEING AMBIDEXTROUS TELL US?

Being ambidextrous—that is, being equally left- and right-handed—is not normal. Nor is it a good thing. The reason? It slows down the brain.

In order for the brain to operate at its peak complexity, each side of the brain must be specialized. If both sides of the brain can handle a task equally well, then the brain is not at its most efficient. Its ability to function slows.

Each side of the brain for the most part controls the opposite side of the body for voluntary muscle (motor) activity. The left brain is dominant for voluntary motor control, which is why most people are right-handed. Even in most left-handed people, the left hemisphere dominates control over this activity. If you are truly ambidextrous, which is actually somewhat rare, it means the left hemisphere is not dominant in this task. Because there is no specialization, both sides are slower. It's like having too many cooks in the kitchen.

Many autistic children, by the way, are left-handed. When they go through the Brain Balance Program, many of them naturally migrate to right-hand dominance. This signifies the brain has matured, and with that maturity comes better specialization.

Handedness is something that shouldn't become apparent until around age two to three, when the part of the brain responsible for this task is maturing. If you notice handedness earlier than that, it is abnormal. It usually means that the nondominant hand is weaker than it should be.

LEFT VERSUS RIGHT

Before getting started on the quiz that will reveal your cognitive style, here is a glimpse into what the left and right brains are all about.

How the World Looks from the Left

The left hemisphere is all about details. It sees the trees rather than the forest. It breaks the world up into small, discrete pieces, and it analyzes every piece step-by-step, looking for a pattern to emerge. It then remembers that pattern and uses it to predict the most likely outcome and plan the most appropriate response.

The left hemisphere loves patterns. It loves to figure out a pattern and repeats a pattern over and over again. It is more interested in things that are man-made and mechanical than biological or natural.

The left hemisphere takes things literally. It likes the singular literal meaning of a word. We talk to ourselves mostly with our left brain. It is logical and linear and is in charge of basic math operations and calculations. The left hemisphere learns consciously, explicitly, declaratively, and deliberately.

The left hemisphere gives us clarity. It allows us to grasp something physically or mentally. It gives us the ability to handle the world around us.

How the World Looks from the Right

The right hemisphere is all about the big picture, the gestalt or the gist of things. It processes all its information at once. There is nothing logical or linear about the right brain. It takes a holistic view of the world.

The right hemisphere rules our posture, gait, and balance and most of our subconscious movements. If you're a great dancer, you have your right brain to thank for that. It's also in charge of nonverbal communication. You read body language with your right brain. It also gives you the rhythm and intonation to sing.

The right brain helps us to read other people's intentions and emotions. It also allows us to be in touch with our own body and our own emotions. It's where we get our gut feelings.

THINKING BACK

How far back can you remember? No matter how good your memory is, no one remembers walking or crawling around in diapers. There's a reason for this. The area of the brain that houses long-term memory doesn't develop until we are around age three, when brain-building primarily switches to the left brain. Most children can't remember before age three and some won't hone the skill until age seven.

The right brain comprehends that the world is always changing and that nothing stays the same forever. It understands metaphor, symbolism, humor, and music. Imagination is stored here.

Our subconscious and intuition reside in the right hemisphere. It guides us in knowing right from wrong. It houses the rules of social behavior and what society deems as appropriate behavior. It lets us know what others think about us and allows us to care deeply about wanting to be liked and accepted. If our right brain is overactive we can become paranoid that people don't like us or are plotting against us.

Left Brain	Right Brain
Focus on details	Focus on the big picture
Small muscle control	Large muscle control
Verbal language	Body language
Literal	Figurative
IQ	EQ
Approach behavior	Avoidance behavior
Positive emotions	Negative emotions
Math calculation	Math reasoning

Left Brain	Right Brain
Word reading	Reading comprehension
Explicit memory	Implicit memory
Practical, deliberate	Unpredictable
Curious	Cautious
Activates the immune system	Suppresses the immune system
Pro-inflammatory immune system	Anti-inflammatory immune system
Top-down control	Bottom-up control

THE MELILLO COGNITIVE STYLE ASSESSMENT

This assessment will help you determine your cognitive style—that is, whether your tendency is to be more right-brained or more left-brained. Choose the response that best describes your natural tendency, not your learned behaviors. Think about yourself as a child, teenager, or young adult and how you would have answered back then. It is very important that you choose one answer to each question, even if you don't think it fits you exactly. Do not leave any blanks!

1. ☐ A. I like to do and learn things one step at a time.
    ☐ B. I like to do and learn many things at the same time.

2. ☐ A. I tend to focus on details.
    ☐ B. I tend to focus on the big picture.

3. ☐ A. I don't always get the joke or think something is as funny as others.
    ☐ B. I always get the joke, even before others.

4. ☐ A. I don't like change.
   ☐ B. I need to change things often.

5. ☐ A. I like routines.
   ☐ B. I rarely do anything the same way twice.

6. ☐ A. I have very good handwriting.
   ☐ B. I have poor handwriting.

7. ☐ A. I like when things are clearly spelled out and precise.
   ☐ B. I like to think in generalities.

8. ☐ A. I tend to take things literally.
   ☐ B. I am good at reading between the lines.

9. ☐ A. I will read a contract or instructions over and over to make sure I don't miss anything.
   ☐ B. I don't like reading contracts or instructions.

10. ☐ A. I believe or have been told I have a high IQ.
    ☐ B. I believe or have been told I have an average IQ.

11. ☐ A. I did better on the math portion of the SAT.
    ☐ B. I did better on the verbal portion of the SAT.

12. ☐ A. I liked school and I am good at academics.
    ☐ B. I didn't like school and it affected my grades.

13. ☐ A. I am good at learning by rote memorization and repetition.
    ☐ B. I learn best by just doing something.

14. ☐ A. I would prefer to work with computers.
    ☐ B. I would prefer to work with people.

15. ☐ A. I am not good at new ideas.
    ☐ B. I am very good at coming up with new ideas.

16. ☐ A. I am not good at creative problem solving.
    ☐ B. I am very good at problem solving, especially when it takes a creative solution.

17. ☐ A. I was better at algebra than geometry in school.
    ☐ B. I was better at geometry than algebra in school.

18. ☐ A. It is easy for me to visualize things.
    ☐ B. It is hard for me to visualize things.

19. ☐ A. I cannot rotate objects in my mind easily.
    ☐ B. I can rotate objects in my mind easily.

20. ☐ A. I have difficulty making friends.
    ☐ B. I make friends easily.

21. ☐ A. I do not get along with the opposite sex well.
    ☐ B. I get along very well with the opposite sex.

22. ☐ A. I am not an emotional person and never show emotions.
    ☐ B. I am emotional and show emotions easily.

23. ☐ A. I prefer individual sports.
    ☐ B. I prefer team sports.

24. ☐ A. I can never tell what someone is thinking.
    ☐ B. I always think I know what someone is thinking.

25. ☐ A. I like to read.
    ☐ B. I don't read a lot.

26. ☐ A. I am very good at spelling and grammar.
    ☐ B. I am not great at spelling and grammar.

27. ☐ A. I like to read technical and nonfiction books.
    ☐ B. I like to read novels and stories.

28. ☐ A. If I don't understand a word, I will stop and look it up more often than not.
    ☐ B. If I don't understand a word, I generally just move on and figure it out later.

29. ☐ A. I have always been able to do calculations easily in my head.
    ☐ B. I don't do calculations in my head well; I need to write it down.

30. ☐ A. I like numbers; I am good with numbers.
    ☐ B. I don't like numbers.

31. ☐ A. I am more book-smart than street-smart.
    ☐ B. I am more street-smart than book-smart.

32. ☐ A. I like planning ahead.
    ☐ B. I hate to plan; I just want to figure it out as I go.

33. ☐ A. I am not good with metaphors; I like facts.
    ☐ B. I like metaphors or hypothetical examples.

34. ☐ A. I will read the instructions closely before I try something.
    ☐ B. I never read instructions; I prefer to jump in feet first.

35. ☐ A. I sometimes struggle with the main idea of a story.
    ☐ B. I always get the main idea of a story.

36. ☐ A. I am better at understanding than doing.
    ☐ B. I am better at doing than understanding.

37. ☐ A. I am logical; I tend to think things through very carefully before doing.
    ☐ B. I am intuitive; I like to act by "gut instinct."

38. ☐ A. I have a great memory for facts and details.
    ☐ B. I don't have a great memory for facts or details.

39. ☐ A. I remember names, not faces.
    ☐ B. I am very good with faces, but forget names.

40. ☐ A. I have a terrible sense of direction.
    ☐ B. I have a very good sense of direction.

41. ☐ A. I have an explosive anger if I am pushed.
    ☐ B. It takes a lot to get me angry; things don't tend to bother me.

42. ☐ A. I like to work by myself.
    ☐ B. I like to work together as a team.

43. When someone says they have good news and bad news:
    ☐ A. I like to hear bad news first.
    ☐ B. I like to hear good news first.

44. ☐ A. I am good at saving money.
    ☐ B. I am not good at saving money.

45. ☐ A. I like to hold on to things; it takes a lot for me to throw something out.
    ☐ B. I like to get rid of old things and replace them with new things.

46. ☐ A. I like realistic art.
    ☐ B. I like abstract art.

47. ☐ A. I don't really focus on how I look.
    ☐ B. I am very aware of how I look.

48. ☐ A. I don't notice what others think of me.
    ☐ B. I notice and care a lot about what others think of me.

49. ☐ A. I don't know or follow fashion trends.
    ☐ B. I love wearing the latest styles.

50. ☐ A. I prefer to wear classic clothes that I have worn for years and are comfortable.
    ☐ B. I prefer to wear newer, trendier styles even if they are uncomfortable.

51. ☐ A. Some people would consider me a geek.
    ☐ B. No one would ever consider me to be a geek.

52. ☐ A. I generally obey laws and follow the rules.
    ☐ B. I generally don't follow rules; I make up my own rules; most rules don't make sense.

53. ☐ A. I work better with positive reinforcement; I work to achieve a goal.
    ☐ B. I work better with negative reinforcement; I focus on avoiding failure.

54. ☐ A. I am very neat and organized.
    ☐ B. I would be considered messy and disorganized.

55. ☐ A. I like to be alone.
    ☐ B. I like being around others.

56. ☐ A. I never remember the words to a song; I like the music more.
    ☐ B. I like the words to a song and remember them almost instantly.

57. ☐ A. I prefer red, yellow, or orange (warm colors).
    ☐ B. I prefer purple, blue, or green (cool colors).

58. ☐ A. I like things that are man-made and mechanical.
    ☐ B. I like things that are natural.

59. ☐ A. I am a perfectionist.
    ☐ B. I don't care if things are not perfect.

60. ☐ A. I would never write, or show someone something I have written, before checking for grammatical or spelling errors.
    ☐ B. I am more interested in the overall content of something I write rather than the details, such as spelling and grammar.

61. ☐ A. I am not good at creative writing.
    ☐ B. I like to write my own stories.

62. ☐ A. I like to listen to classical music.
    ☐ B. I like popular music (rock or country).

63. ☐ A. I am very good at learning languages.
    ☐ B. I am terrible at languages.

64. ☐ A. I am better at reading books than people.
    ☐ B. I am better at reading people than books.

65. ☐ A. I mentally comprehend suffering, but I don't really feel it.
    ☐ B. I feel very bad or sad for others who are suffering.

66. ☐ A. I rarely get depressed.
    ☐ B. I get depressed easily.

67. ☐ A. I generally don't like to be touched, especially by someone I don't know.
    ☐ B. I need human contact and like to be touched and to touch others.

68. ☐ A. I am somewhat uncoordinated, not very athletic.
    ☐ B. I am generally very coordinated and athletic.

69. ☐ A. I'd rather stay indoors.
    ☐ B. I'd rather be outside.

70. ☐ A. I like to vacation at the same place over and over.
    ☐ B. I like to vacation in new places.

71. ☐ A. I don't like parties and social gatherings in general.
    ☐ B. I love parties and social gatherings.

72. ☐ A. I am a realist.
    ☐ B. I am a dreamer.

73. ☐ A. Function is much more important than style or design.
    ☐ B. Design is at least as important as function.

74. ☐ A. I prefer math, research, or science.
    ☐ B. I prefer philosophy and mythology.

75. ☐ A. I would prefer to communicate through text or email.
    ☐ B. I would prefer to communicate on the phone or in person.

76. ☐ A. I am not a people person.
    ☐ B. I am definitely a people person.

77. ☐ A. I prefer to be organized and plan things.
    ☐ B. I prefer to be spontaneous and not worry about the details.

78. ☐ A. I think it is most important to improve on things that exist and make them better.
    ☐ B. I think it is most important to develop new things and new ideas.

79. ☐ A. I think reason is more important than feelings.
    ☐ B. I think feelings are more important than reason.

80. When learning a new chapter in a textbook:
    ☐ A. I think it is best to outline the chapter.
    ☐ B. I think it is best to summarize the chapter.

81. ☐ A. I am better at crossword puzzles.
    ☐ B. I am better at jigsaw puzzles.

82. In a theater production, I would rather:
    - ☐ **A.** Be the director.
    - ☐ **B.** Be the lead actor.

83. When learning a new piece of equipment I:
    - ☐ **A.** Carefully read the instruction manual before beginning.
    - ☐ **B.** Jump in and wing it (I use the manual as the last resort).

84. ☐ **A.** What is being said (words) is more important than how it is being said (tone, tempo, volume, emotion).
    - ☐ **B.** How something is being said (tone, tempo, volume, emotion) is more important than what the person is saying.

85. ☐ **A.** I do not use hand gestures when I speak.
    - ☐ **B.** I use many gestures and hand movements when I speak.

86. If I were hanging a picture on a wall, I would:
    - ☐ **A.** Carefully measure to be sure it is centered and straight.
    - ☐ **B.** Put it where it looks right and move it if necessary.

87. At work:
    - ☐ **A.** I concentrate on one task at a time until it is complete.
    - ☐ **B.** I usually juggle several things at once.

88. ☐ **A.** I like to plan my future steps.
    - ☐ **B.** I enjoy dreaming about my future.

89. ☐ **A.** I like to take ideas apart and look at them separately.
    - ☐ **B.** I like to put ideas together.

90. ☐ **A.** I like to learn about things we are sure of.
    - ☐ **B.** I like to learn about hidden possibilities.

91. I think it is it is more exciting to:
    - ☐ **A.** Improve something.
    - ☐ **B.** Invent something.

92. I am strong:
    - ☐ A. In recalling verbal materials (names and dates).
    - ☐ B. In recalling spatial material (directions and locations).

93. ☐ A. I prefer total quiet when reading or studying.
    - ☐ B. I prefer to have music on while reading or studying.

94. ☐ A. I think in words.
    - ☐ B. I think in pictures.

95. As a kid, the worst thing would be to:
    - ☐ A. Fail a test.
    - ☐ B. Be embarrassed in class.

96. ☐ A. I learn best from teachers who explain with words.
    - ☐ B. I learn best from teachers who explain with pictures, movement, and actions.

97. ☐ A. I like to express feelings and ideas in plain language.
    - ☐ B. I like to express feelings and ideas in poetry, song, dance, and art.

98. ☐ A. I would rather not guess or play hunches.
    - ☐ B. I like to play hunches and guess.

99. ☐ A. I am very direct and straightforward with people.
    - ☐ B. I try not to hurt someone's feelings so I am not as direct with people.

100. ☐ A. I think the best trait is to be reserved and modest.
    - ☐ B. I think the best trait is to be outgoing and interesting.

SCORING

To find out your cognitive style, add up all the As and Bs. Subtract the lower score from the top score. Then plot the resulting

number on the A or B side of the scale. A is left-brain dominant; B is right-brain dominant.

Example: 80 As − 20 Bs = 60 A

100 A   60 A   0   100 B

A COUPLE'S SCORE

To calculate the overall score of you and your partner as a
couple: If you are both left- or right-brain dominant, average
your two numbers to get an overall couple score. If one of you
is right-brain dominant and the other is left-brain dominant,
subtract the lower score from the higher one to get an overall
couple score.

Interpreting the Results

Individual Assessment

If you scored between 0 and 24 for left or right brain: You have a mild
laterality, meaning your cognitive style is fairly mixed. There is not a
big difference in the way you use both sides of your brain. Regardless
of whether you are more right-brain dominant or left-brain dominant, your risk of having autistic traits is very low.

If you scored between 25 and 49 for left or right brain: You have a
moderate laterality and are more obviously dominant on one side or
the other. If you are more right-brain dominant, your risk of having
any autism traits is very low. If you are left-brain dominant, your
chances of passing on autism traits is a bit higher than the previous
group but still relatively low.

If you scored 50 to 74 for right brain: You have strong right-brain
laterality and you are right-brained. Your risk of having or passing on
autistic traits is very low.

If you scored 50 to 74 for left brain: You have strong left-brain laterality and are left-brained. Your risk of having and passing on autistic traits is high.

If you scored 75 or above for right brain: You are considered to have extreme laterality and you are extremely right-brained. The likelihood of you having autism traits is extremely low.

If you scored 75 or above for left brain: You are also considered to have extreme laterality and are extremely left-brained. You are at highest risk of having or passing on autism traits.

Couple's Assessment

The combination of traits inherited from both parents ultimately determines your child's cognitive style. Of course, this is not an exact science. How the mixture of parental traits is passed on is unknown, but I believe there is a measure of predictability. For example, if both the mother and the father have cognitive styles predictive of autistic traits, then the child's risk of having autistic traits is high. If both parents are more right-brained, then the risk is very, very low. If they are a mixture—one is right-brained and the other is left-brained—then the risk is somewhat iffy. However, in general, the risk of passing on these traits to a child is highest if both parents are left-brained.

To assess this, combine your two scores, then calculate your average and see where you fall as a couple on the dominance scale. The risk for a couple is basically the same as for an individual.

PROFILE OF A TYPICAL LEFT-BRAIN PERSONALITY

The left-brain personality generally craves an orderly life. Predictability and familiarity are their operative words. They have habits and routines they like to follow daily, and they can get out of sorts when someone comes along who wants to shake things up. They have a tendency to be obsessive or compulsive when matters get out of control.

Left-brain cognitive types are rule followers. When they buy something new, they open the box and read the instructions. They are good with details and like numbers. They generally are not spontaneous, although they can be impulsive. They generally are happy people, especially when they are young, but they can have a tendency to become anxious as they get older. They worry easily and can get agitated.

A typical left-brained person's creative interests lean more toward classical music. They may also be good at music themselves, though their ability generally does not go beyond playing an instrument. You most likely won't hear them belting out a song. They are also good with their hands and have nice handwriting. They can make small, detailed letters. This also makes them good at art, and they are likely to take up painting at some point in life.

When they communicate, they like getting to the point. They are not one for projecting emotion and using descriptive language. They were playful and outgoing as kids, but tend to get less so as they grow into adulthood. In school they were more interested in academics than dating. Social situations make them uncomfortable. They tend to hate crowds.

You don't find left-brain people making rash decisions and acting on their hunches. When they do, they feel out of their element. In fact, they often have a hard time making a decision because they get so caught up in the details. They like to analyze everything. So, it is no surprise that they have a very good memory for details. They do things one step at a time; they do not like to multitask.

Left-brainers can have some odd affectations. They tend to be quite set in their ways and generally are not very flexible. They're the type that always have to be doing something and can't sit still. They are hyperactive and very distractible. In conversation, they can get on a new subject before finishing another.

LEFT-BRAIN GENIUS: ISAAC NEWTON

The perfect example of a left-brain genius was Sir Isaac Newton, who is considered by many to be the greatest scientist who ever lived. Personally, he was not well liked and considered arrogant by many. Reportedly, he never had a romantic relationship with anyone his entire life.

By all historical accounts, Newton had autistic traits. Though brilliant, he was obsessive and compulsive, especially about numbers and patterns. He was socially inept and did not like being around or talking to people.

I suspect that if Isaac Newton were around today and was exposed to modern social and environmental factors, he would probably be diagnosed with autism and perhaps function at a much lower level than he did in his day, all with the exact same genetic profile.

As neat and orderly as they are about their personal space, they are not the same in their personal style. They are unaware of social norms, especially when it comes to dress. Fashion bores them and they never change their hairstyles. They are not usually good at many sports, so you won't find them on the office volleyball team. However, you will find them glued to their computers and cell phones because they find technology fascinating. They generally don't have a lot of friends, especially of the opposite sex.

Left-brainers are realists. They don't like joking around and often don't even get a joke. They generally don't read novels and stick to literature and nonfiction. They are more book-smart rather than street-smart. They are more theoretical than practical.

People with a left-brain cognitive style gravitate toward professions in which they can work alone or not interact a lot with people. Because they are so good at small details, you might find them making

THE MIND OF A GENIUS

The brain breeds two types of geniuses. One, which I called the balanced genius, comes from an incredibly balanced and integrated mind. It is a brain so finely tuned and perfectly balanced that it works as a whole seamlessly, easily, and flexibly. This person has a strong left brain *and* a strong right brain and can use it in a way few people can.

The best example of this and the person considered the greatest genius of all time was Leonardo da Vinci. In fact, he trained himself from a young age to have a perfectly balanced brain. He trained himself to write and paint equally well with his left and right hands. He could write right side up and upside down, forward and backward, all with apparently equal skill. He was a master of fine art, but he was also inventive and abstract. He created new inventions and new techniques in painting. But yet he also could draw and copy detail perfectly. He had the perfect balance of science and art. He was a balanced genius.

The unbalanced genius comes from a brain with a disability. This is the person who is incredibly gifted at certain abilities, but also clearly deficient in others. This is what we see with savant syndrome, or autistic savants. This is a mind with left-hemisphere skills that are clearly exceptional. They may excel in academics, have amazing math skills, or memory for details and dates. They may be great at learning, playing, or remembering music. However, they are equally as bad at social skills and nonverbal communication; they can't read people or faces; they don't hear tone of voice.

Left-brain gifted children are generally considered gifted in school and early academic pursuits. They may be early word readers to the point where they can spontaneously read or write with almost no instruction at a very young age. Many of these children, however, eventually struggle with social skills, and even though they seemed gifted or even genius-like when they are young, they often end up struggling in school and life later on, especially with relationships.

jewelry or working on an assembly line. These are among typical left-brain life jobs:

Accountant	Doctor
Actuary	Engineer
Analyst	Professor
Computer and information technologist	Researcher
	Surgeon

Profile of a Typical Right-Brain Personality

Right-brain cognitive types tend to be our leaders. This is because they are "big picture" people. They see the world as ever-changing and unpredictable and they like it that way. Routine bores them and they are always exploring for something new to try.

Right-brain people can be just as smart as the next person, but they don't necessarily excel in school. In school they were quite popular and enjoyed being the center of attention. Despite having excellent social skills, however, they still can also be painfully shy. They really care what others think, which is part of their fully loaded emotional makeup. They are emotional and empathetic, and can feel another's pain. They feel negative emotions strongly and, as a result, can get depressed easily. The people who you can hear sobbing in the movie theater are doing it with their right brain.

Right-brainers tend to be very spatial. They like to move, but they are more sensory in nature. They learn more subconsciously and non-verbally. They are good at reading facial expressions and body language. Others would say they have good intuition.

Right-brainers also make good athletes because the right brain controls the large muscles. They have good body control and balance.

They are practical, not theoretical. When they rip open a box, they start assembling immediately. To them, directions just get in the way and they usually can figure something out pretty quickly. They learn more by doing and interacting. This makes them more street-smart than book-smart.

Right-brain cognitive types can be charismatic. They are great communicators. They like participating in team sports and joining clubs. They are creative in big ways. They are idea people and inventive. They can create things in their minds. They get and like abstract art. Musically, they tend to be singers or songwriters rather than a player in the band.

Despite their potential for great things, they can struggle with motivation and have a tendency to be lazy. To this end, they generally are not neat by nature. They let their house and office get a mess, even though aesthetics are important to them. However, they like dressing well and keeping up with the latest fashions. Their reading habits lean toward novels.

RIGHT-BRAIN GENIUS: ALBERT EINSTEIN

A perfect example of a right-hemisphere genius was Albert Einstein. Talk about being able to see the big picture. He was able to see the whole universe better than anyone in history!

For all his genius, Einstein was a poor student all the way through postgraduate school. He barely finished his PhD and he was fourth out of five in his graduating class. He was not a great mathematician by nature, but he was great at the concepts of physics. He would do what he called thought experiments where he would come up with big ideas, and then he would figure out the math to support it.

He was a likable guy and reportedly had many romantic relationships. He loved people and he liked the limelight.

Right-brain personality types are funny and engaging. They are pragmatic and good at metaphors and humor. Right-brain types who can motivate themselves and can stay focused on a goal can be very successful. They gravitate toward jobs that put them at the center of attention or around people. Steve Jobs is a good example of a right-brained person. These are among typical right-brain types of jobs:

Architect	Mechanic
Artist	News reporter
Athlete	Novelist
Builder	Politician
Carpenter	Publicist
CEO	Salesperson
Entertainer	Singer/songwriter
Inventor	Teacher
Mason	

CHAPTER SIX

Measuring for a Brain Imbalance

You now know your cognitive style. If your score was in the strong to extreme range on either hemisphere—that is, 50 to 100—it is an indication that you are better than average at certain things. It also means that you are most prone to developing a brain imbalance.

Having a brain imbalance is not necessarily a bad thing. We're all at least a little out of balance. Virtually nobody has a perfectly balanced brain. If we did we'd be, well, perfect. What you want to know is how mild or severe your imbalance is. That's what taking the Melillo Adult Hemispheric Checklist is all about.

This questionnaire is modeled after the checklist we use at our Brain Balance Achievement Centers to identify a brain imbalance in children and assess where the imbalance exists.

It is different than looking at your cognitive style. In that profile we focused on identifying your strengths. This checklist will help you identify any weaknesses that could be causing an imbalance and if they are mild or extreme. A few weaknesses in and of themselves are not a problem. Too many, however, can impair the way you function through your daily life. The more weaknesses you have, the more of an imbalance, the greater the impairment, and the greater the risk of passing them on to your children. However, this needn't happen. A brain imbalance can be corrected. There is remedial action for it, which is what the rest of this book is about.

MELILLO ADULT HEMISPHERIC CHECKLIST

This assessment consists of two hundred characteristics of brain imbalance—one hundred are associated with a right-brain deficit and one hundred are associated with a left-brain deficit. They are divided into seven categories, not necessarily equal in number. Go through all, both left and right, and check off all that apply to you. At the end I will tell you how to tally your score and how to interpret it.

If you find you have an imbalance, this checklist is something you'll want to come back to from time to time. When you do so, note if the symptom is improving or getting worse.

Motor Characteristics of a Right-Brain Deficit

- ☐ 1. **You have poor coordination. You are clumsy and have an odd posture and gait.**

- ☐ 2. **You have difficulty coordinating both sides of your body, such as in running, walking, or swimming.**

- ☐ 3. **You have poor muscle tone. Your muscles are flabby or floppy instead of taut.**

- ☐ 4. **You are very flexible, or even double-jointed.**

☐ 5. You have, and may recently have acquired, facial tics or make repetitive and involuntary vocal sounds, such as clearing the throat.

☐ 6. You have repetitive motor mannerisms, such as twisting or playing with your hair or pulling on your sleeves.

☐ 7. You have a tendency to walk on your toes.

☐ 8. You do not have good balance or notice it has recently gotten worse.

☐ 9. You have chronic tendinitis, patella tendinitis, carpal tunnel syndrome, or have or have had trouble with your rotator cuff or tennis elbow on the right side of the body.

☐ 10. You get a tremor or eye twitch on the right side of your body.

☐ 11. You cannot cross your eyes.

☐ 12. You have poor depth perception. For example, you have trouble judging distances or driving.

_____ Total (A)

Motor Characteristics of a Left-Brain Deficit

☐ 1. You have difficulty performing fine motor skills, such as buttoning a shirt and doing small detail work with hands.

☐ 2. You have poor handwriting or handwriting that is hard to read.

☐ 3. You have difficulty planning a sequence of coordinated movements, such as dance steps or certain sports activities.

☐ 4. You have a twitch in your right eye, but not your left.

☐ 5. You frequently have problems with writer's cramp.

☐ 6. You stumble over words in your speech when you're tired.

☐ 7. You are not musically inclined. For example, you have found it difficult to play a musical instrument.

☐ 8. You have chronic tendinitis, patella tendinitis, carpal tunnel syndrome, or have had rotator cuff problems or tennis elbow on the left side of the body.

☐ 9. You get tremors on the right side of the body.

☐ 10. You find it difficult to imitate an action without actually doing it. For example, you can't mimic how to strike a match without using a pack of matches.

☐ 11. You're ambidextrous. You don't really favor your left hand or right hand.

_____ Total (B)

Sensory Characteristics of Right-Brain Deficit

☐ 1. You have poor spatial orientation. For example, you frequently bump into things.

☐ 2. You are hypersensitive to sound. For example, you hate loud noises, such as fireworks, and you are especially bothered by high-pitched sounds, such as children screaming or scratching on a chalkboard.

☐ 3. You generally feel disconnected from your body.

☐ 4. You compulsively touch things. For example, you'll touch fabric when passing through an aisle in a store even though you're not interested in buying, or you rub your hands over furniture for no particular reason when you're in someone's home.

☐ 5. You don't like the feeling of clothing on your arms or legs. For example, you'll pull off clothes at every chance you get.

☐ 6. You don't like being touched or when people get in your personal space.

☐ 7. You have a poor sense of smell. For example, you don't feel a hit when you get in an area of a wood fire burning, popcorn popping, or cookies being baked in the oven.

☐ 8. You have an inability to recognize or differentiate between sounds or musical notes. You may have been told you have a "tin ear" or you sing off-key.

☐ 9. You have experienced hearing voices when no one is there or hear a ringing in your right ear.

☐ 10. You smell unusual scents others don't smell or when none exists.

☐ 11. You get a metallic or unpleasant taste in your mouth for no discernable reason, such as a side effect from taking certain medications.

☐ 12. You are obsessed with religion or understanding the meaning of life, beyond what many would consider normal. You are always quoting scripture in almost every conversation or Facebook post.

☐   13. You have unexplained lapses in time. You don't lose consciousness, but you can't remember certain periods of time.

☐   14. Your right eye is chronically irritated, dry, or red.

☐   15. Hearing in your left ear is more difficult than in your right.

☐   16. You seem to lose your perception of time, or you always had a poor sense of timing, meaning you over- or underestimate how much time something will take.

_____ Total (A)

Sensory Characteristics of Left-Brain Deficit

☐   1. You have to think twice when distinguishing left from right.

☐   2. You or others feel you do not hear well, even though hearing tests have come out normal.

☐   3. You easily get motion sickness.

☐   4. You are very sensitive to movement. For example, you can't ride in a car and read at the same time.

☐   5. The smallest things can make you nauseated, such as the sight of blood or certain unpleasant smells.

☐   6. You feel you don't hear as well with your right ear as you do with your left.

☐   7. Your left eye is chronically irritated, dry, or red.

_____ Total (B)

Emotional Symptoms of Right-Brain Deficit

☐ 1. You are, or appear to others to be, overly happy and affectionate. For example, you love to hug and kiss others, such as your kids, friends, and pets.

☐ 2. Your behavior could be described as manic. You can burst into tears or laughter almost spontaneously.

☐ 3. You have sudden outbursts of anger or fear.

☐ 4. You occasionally experience panic and/or anxiety attacks.

☐ 5. You sometimes have dark or violent thoughts.

☐ 6. You hold on to past "hurts." You just can't let go.

☐ 7. Your face normally lacks expression and you don't exhibit much body language when interacting with others.

☐ 8. You're too uptight. You just cannot seem to loosen up.

☐ 9. You are not very empathetic or do not appear to be to others.

☐ 10. You don't show emotion in situations the same as others do.

☐ 11. You like taking risks and are known as a risk taker.

☐ 12. Normally, you speak in a monotone and have no expression in your voice.

☐ 13. You don't like being in social situations. Others would call you antisocial.

☐ 14. You don't generally ever feel "afraid" or seem to experience fear.

_____ Total (A)

Emotional Characteristics of Left-Brain Deficit

☐ 1. You get frightened very easily.

☐ 2. You frequently and easily get depressed or feel down in the dumps.

☐ 3. You worry a lot and are considered a worrywart by those who know you best.

☐ 4. You have had or think you have had post-traumatic stress disorder.

☐ 5. You have a lot of fears and/or phobias.

☐ 6. You frequently get moody and irritable.

☐ 7. You contemplate suicide.

☐ 8. You lack motivation.

☐ 9. You don't get a lot of pleasure out of life, food, or anything in particular.

☐ 10. Others would say it is hard to make you happy.

☐ 11. You get insulted easily.

☐ 12. You frequently feel overwhelmed by the tasks at hand and what's going on in the world around you.

☐ 13. You can feel another's pain and despair more than the average person.

☐ 14. You're typically pessimistic. Others would call you extremely negative.

☐ 15. You are excessively cautious and don't take risks.

☐ 16. You're uncomfortable in social situations. You want to be sociable but you are not always sure how to act.

☐ 17. Your feelings are hurt easily and it can make you cry at the drop of a hat.

☐ 18. You sometimes have feelings of hopelessness, or feel "What's the point?"

☐ 19. You are very sensitive to what others think about you.

☐ 20. You are overly self-conscious. Others might call you paranoid.

☐ 21. You often think that others are making fun of you behind your back.

☐ 22. You have bad memories that you just can't let go of, especially emotional hurt or humiliation.

_____ Total (B)

Behavioral Characteristics of Right-Brain Deficit

☐ 1. You have a hard time following the rules of good communication. You say inappropriate things, talk at the wrong time, and you are not particularly expressive when talking.

☐ 2. You have, or others tell you that you have, a hard time staying focused on the task at hand or paying attention to what is being said or done around you for more than a few minutes.

☐ 3. You sometimes think you have or others accuse you of having adult ADHD. Or you have been diagnosed with it.

☐ 4. You obsess over practically everything. You often think or others have told you that you act like you have obsessive-compulsive disorder or you have been diagnosed with it.

☐ 5. You have manic-depressive highs and lows to the degree that you or others feel you might have bipolar disorder or you have been diagnosed with the disorder.

☐ 6. You frequently have feelings of déjà vu. You feel like you have been somewhere or experienced an event before.

☐ 7. You often miss the point of a story.

☐ 8. You get stuck in set behavior and can't let it go. For example, you can't transition easily to a new thought, action, or idea.

☐ 9. You never feel a sense of guilt or remorse.

☐ 10. You lack social tact and feel socially isolated.

☐ 11. You manage your time poorly. You are always late for appointments and meetings.

☐ 12. You are a neat freak and can't stand when something is out of order.

☐ 13. You can't sit still. You are impulsive, compulsive, and hyperactive.

☐ 14. You have a hard time getting to sleep because your mind is always racing.

☐ 15. You hate throwing things out and may even be accused of being a hoarder.

☐ 16. You're generally uncooperative.

☐ 17. You have extreme eating habits, possibly to the point of an eating disorder, especially bulimia.

☐ 18. You often appear to others as bored, aloof, and abrupt.

☐ 19. You are considered strange by others or you were considered strange as a kid.

☐ 20. You don't have a lot of friends. New friends you make seem to drift away.

☐ 21. You don't particularly enjoy the company of others.

☐ 22. You act silly or giddy at inappropriate times in the presence of others.

☐ 23. You talk incessantly and are known to others as "a talker."

☐ 24. You tend to ask or have been told you ask repetitive questions and talk in circles, but never really get to the point.

☐ 25. You couldn't care less about fashion and social trends.

☐ 26. You've been wearing the same hairstyle for years.

☐ 27. You have been described as a "control freak."

_____ Total (A)

Behavioral Characteristics of a Left-Brain Deficit

☐ 1. You tend to be oblivious to rules and regulations. For example, you seem to "attract" parking and speeding tickets.

☐ 2. You have a tendency to exaggerate and/or lie.

☐ 3. You can feel terrible shame or crippling guilt even when you have not done anything terribly wrong.

☐ 4. You frequently have a foreboding feeling or feel sick in your stomach for no discernible reason.

☐ 5. You tend to procrastinate.

☐ 6. You are very shy, especially around strangers.

☐ 7. You have a tendency to stutter or stammer.

☐ 8. You have poor self-esteem. You feel like a loser and you feel others think of you that way.

☐ 9. You have, or others say you have, an inferiority complex. You don't feel you're as smart as others.

☐ 10. You hate busywork, such as doing paperwork or paying the bills.

☐ 11. You are not good at following routines and establishing habits.

☐ 12. You get perplexed, frustrated, or annoyed when you have to follow multiple-step directions.

☐ 13. You tend to jump to conclusions or have been told you jump to conclusions too quickly.

☐ 14. You make mistakes because you don't read or follow directions.

☐ 15. You are extremely messy.

☐ 16. You follow fads and trends and are beholden to the latest fashions.

☐ 17. You were bored in school and/or hated schoolwork.

☐ 18. You just hate being alone.

☐ 19. You are a daydreamer and are constantly drifting off in thought.

☐ 20. You have a hard time doing just one thing at a time.

☐ 21. You constantly stop one activity or thought and start another before anything is finished.

☐ 22. You have an addictive personality. For example, you have been addicted to drugs, alcohol, cigarettes, gambling, sex, etc.

_____ Total (B)

Cognitive Characteristics of Right-Brain Deficit

☐ 1. You have difficulty recalling the date, time, or place of important events in your life. For example, you don't just forget birthdays and anniversaries, you sometimes have to pause to remember your age or day and year you were married.

☐ 2. You have trouble staying on topic during a conversation or at a meeting.

☐ 3. You have trouble repeating a story as told or explaining directions.

☐ 4. You have a tendency to fly off the handle to common annoyances, such as an overflowing sink or an overheated car engine.

☐ 5. You have difficulty interpreting abstract language. For example, you may not get a joke or understand a metaphor.

☐ 6. You cannot mentally rotate objects in space. For example, you have a hard time imagining how furniture would look in a different position of a room.

☐ 7. You can't seem to recognize faces of people you know who you haven't seen in years.

☐ 8. You do not remember much or almost anything about your childhood.

☐ 9. You have a poor sense of direction.

☐ 10. You have difficulty using and understanding innuendo and connotation. For example, you can't take a hint.

☐ 11. You don't "get it" when someone uses irony and sarcasm.

☐ 12. You don't get the moral in a story or the point an author is trying to make.

☐ 13. You get so stuck in the details that you have difficulty making decisions.

☐ 14. You have a hard time making decisions because you tend to overanalyze everything.

☐ 15. You have an obsessive interest in unusual topics, such as trains, rocks, stamps, comic books, and the like.

☐ 16. You are frequently and are growing increasingly impatient.

☐ 17. You speak aloud what's on your mind.

☐ 18. You get very close to people when speaking to them. Others would call you a space invader.

☐ 19. You are extremely direct to the point of being considered rude.

_____ Total (A)

Cognitive Characteristics of Left-Brain Deficit

☐ 1. You have a problem remembering details, such as street names or important dates.

☐ 2. Colors look dull to you lately.

☐ 3. You have trouble remembering names and phone numbers.

☐ 4. You find it difficult to learn by reading.

☐ 5. You have difficulty executing a plan, such as following a recipe or building a model.

☐ 6. You have poor analytical skills. For example, you can't think logically in a stressful situation or analyze your odds of winning or losing in a game.

☐ 7. You have no sense of time. You are always late.

☐ 8. You have trouble prioritizing. For example, you have difficulty knowing what to do first or what is most important.

☐ 9. It is unlikely you would take the time to read the instruction manual before trying something new.

☐ 10. You have to tendency to miss small words when reading or omit them when you write.

☐ 11. You have difficulty learning new material and your reading is too slow and laborious.

☐ 12. Names and words get caught on the tip of your tongue.

☐ 13. You need to hear or see concepts many times in order to learn them.

☐ 14. You believe you are or were dyslexic.

☐ 15. When you were in school, your test scores and grades tended to get worse instead of better.

☐ 16. You took special education classes in grade school or high school.

☐ 17. You say you are poor at math.

☐ 18. You're a bad speller.

☐ 19. You are not particularly good at grammatically correct writing.

☐ 20. You studied a foreign language but can't recall it or can barely recall it.

☐ 21. You have difficulty or can't describe the nature of your relationships in emotional terms, such as what your relationship was with you mother when you were growing up.

☐ 22. You can't remember details of your childhood, such as the addresses where you lived, your phone numbers, or your teachers' names.

_____ Total (B)

Common Immune Characteristics of Right-Brain Deficit

☐ 1. You have allergies.

☐ 2. You have a sensitivity to a food substance, such as casein or gluten.

☐ 3. You have or have had an autoimmune disorder such as asthma, eczema, lupus, psoriasis, or rheumatoid arthritis.

☐ 4. You have more than one autoimmune disorder.

☐ 5. You have little white bumps on your skin, especially on the back of your arms.

☐ 6. You crave certain foods, especially dairy and wheat products.

☐ 7. You have been diagnosed with low thyroid function.

☐ 8. You feel like you're a little drunk or feel off balance after eating certain foods.

_____ Total (A)

Common Immune Characteristics of Left-Brain Deficit

☐ 1. You have problems with chronic ear, throat, or respiratory infections.

☐ 2. You are prone to benign tumors and/or cysts or you have had a cancerous tumor.

☐ 3. You've taken or frequently take a lot of antibiotics or antiviral medicines.

☐ 4. You catch a lot of colds, more than two a year.

☐ 5. It takes you a long time to feel 100 percent after an illness.

☐ 6. You feel you have to get a flu shot every year or you will get the flu. You sometimes get it, even with a flu shot.

☐ 7. You have a problem with chronic yeast or fungal infections and/or have been diagnosed with candidiasis or thrush.

☐ 8. You have or have had stomach ulcers.

☐ 9. You've had pneumonia within the past seven years.

☐ 10. You have recurrent viral outbreaks, such as herpes or shingles.

☐ 11. You have had or still have Lyme disease.

☐ 12. You've had your tonsils and adenoids removed because of chronic infections.

_____ Total (B)

Common Metabolic Characteristics of Right-Brain Deficit

☐ 1. You have frequent bowel troubles with constipation and/or diarrhea.

☐ 2. You have a rapid heart rate or a sudden increase in heart rate (tachycardia, above ninety beats per minute).

☐ 3. Your blood pressure is ten points or more higher when taken on your right arm than your left arm.

☐ 4. You perspire more on the right side of your body than your left.

_____ Total (A)

Common Metabolic Characteristics of Left-Brain Deficit

☐ 1. Your blood pressure is ten points or more higher when taken on your left arm than your right arm.

☐ 2. You sweat more on the left side of your body.

☐ 3. You have or have had an irregular heartbeat, such as arrhythmia or a heart murmur.

☐ 4. Your left hand loses circulation and takes longer to warm up when exposed to the cold.

_____ Total (B)

HOW TO SCORE
Tally the number of checkmarks you made in the right-brain list of deficit symptoms (A) and left-brain deficit symptoms (B). The side with the highest number is the side of hemispheric weakness. The greater the number and the more they are different, the more severe the imbalance between the two sides.

_____ Total number of checkmarks for
right-brain deficit (A)

_____ Total number of checkmarks for
left-brain deficit (B)

_____ Hemispheric weakness right or left

Interpreting Your Results

The more total checks you have, the more likely you are to be diagnosed or suffer with various types of issues involving your health, your relationships, and your mental and emotional state. This largely depends on which specific sections you have made the most checkmarks. For example, if you have more checkmarks for both A and B for immune or metabolic problems relative to other parts, the more likely you are to have an immune or metabolic disorder. However, it is probably not primarily the result of a brain imbalance. On the other hand, if you have a majority of checkmarks in one or several sections on just one side of the brain, then it is likely you have a brain imbalance and the imbalance is centered in the areas where you have most of the checkmarks.

There is no standard number of checkmarks that draws the line between a brain that is balanced and one that is not in balance. Everyone's brain is different. Ideally, you want to have a relatively low number of checks as a whole and the checkmarks that you have should not be significantly more on one side or the other. If you have a particular problem such as depression, for example, taking this test should help you to understand that the cause of the problem may be a brain imbalance. Also, other symptoms and health issues you have may all be related to your brain imbalance.

The more symptoms you have, the more severe your brain imbalance and the more likely you are to have a child who will end up with similar symptoms and a brain imbalance. The scoring is relative based on each individual's health, personality, and genetic makeup.

If your score indicates a brain imbalance, there is no need for alarm. In the next chapter, I will help you identify your brain imbalance issues more specifically and if they need to be addressed. Then in Chapter 11, I offer exercises that will help you improve the balance between the two hemispheres.

CHAPTER SEVEN

Listen to Your Body

Assessing Your Symptoms

You now know your cognitive style and have taken the Melillo Adult Hemispheric Checklist assessment. You have read all about the possible risks for autism in Chapter 4. By now you should be starting to get a pretty good idea just where you and your mate stand in terms of your opportunity to bring healthy children into this world.

If you're like most people, you are surprised to learn how important the health of your brain is to your overall health, especially your reproductive system. Since virtually no brain is perfectly balanced, you are wondering just how out of balance your brain is: not so much that it matters, or just the opposite? This series of self-assessment tests will help you figure that out.

We all carry around with us symptoms of our individual brain imbalance. Many of us literally put our symptoms on display. The signs can be quite obvious; you just have to be looking for them. I am struck all the time at how unaware people are of their own imbalances. I work with people like this every day, so their imbalances

ENLISTING THE HELP OF A PROFESSIONAL

If you think you may have a brain imbalance or any of the related problems discussed in this book, I suggest you get an examination from a specialist in functional neurology. The brain and biological processes discussed in this book are much too complicated for anyone to self-diagnose a specific neurological or metabolic problem. There is no way for you to know for sure if you have a hormone imbalance, a leaky gut, if your stress response is too strong and for how long it's been affecting you, or how to measure if you are sensitive to heavy metals. A functional neurologist or specialist in functional medicine can help you work through it all.

Functional neurology is a complementary medicine based on the philosophy that brain function and body processes are not totally integrated unless the brain is working as a whole. A functional neurologist will put you through a series of noninvasive tests to help pinpoint your physical problems.

I recommend that anyone who believes they have a brain imbalance after reading this book should have a professional evaluation before conceiving.

pretty much jump out at me. One night not too long ago my wife and I were watching the evening network news. A handsome young reporter was filling in as the anchor, but a close look was telling me a completely different story about his looks. *Do you see the imbalance in his face?* I asked my wife. It was so obvious to me. One eye was bigger than the other. His mouth drooped on one corner when he smiled. The crease that runs from the nose to the mouth was deep on one side and barely detectable on the other. His head was tilted to one side and, when he spoke, only one side of his face seemed to exude expression.

The facial features detected in this sketch are indicative of a brain imbalance.

Most people wouldn't notice such a thing and I bet the reporter's bosses hadn't. I took a picture of the young man with my cell phone and his face looked even more out of sync. The well-disguised sketch of that photo that you see here shows symptoms of a brain imbalance. When I show this sketch to students, they guess that he might have had a stroke, though I doubt it very much. But I'd bet he has a lot of health complaints, along with the gifts that have made him a prominent major network news anchor and reporter.

A DOMINO EFFECT

A brain imbalance is responsible for a myriad of health issues and annoyances that millions of people go through daily. All you have to do is turn on the television on any given evening and the proof is in all the commercials you see—acid reflux, erectile dysfunction, high blood pressure, high cholesterol, allergies, dry eyes, dry mouth, sleeping problems, depression, anxiety, memory problems, you name it. These health complaints each have their individual cause, but any functional neurologist will tell you that they all are in some way related to the way your brain is steering your body.

A problem in the brain creates problems elsewhere. This is especially true if you have multiple symptoms. The short list includes blood pressure that is higher when taken on both arms, muscle weakness that is more pronounced on one side of the body, and a decrease in your sense of smell, hearing, or touch on one side relative to the other. You may have very poor balance and/or coordination, get dizzy for no obvious reason, or get extremely depressed, sad, or manic when something unexpected happens. You might start to notice you have trouble staying focused, or you're not as comfortable around people as you used to be. You might become obsessive, compulsive, or fearful when you were never this way before. Headaches and back pain may start to come on regularly. These are problems that used to be associated with getting older, but we are now seeing them in people at a younger and younger age.

One health problem is not necessarily a sign that you have a brain that's out of balance. But the more health problems you accumulate, the more likely your chances are. These quick self-assessment exercises will help you figure it out.

You can do most of these tests on your own, but they are best done in conjunction with a partner, as it can be hard to examine yourself objectively. Enlisting the help of your partner or someone who knows you well can be very helpful. Either way, you will need a large mirror. A video or still camera are useful as well. Also, have a notebook and pen on hand as you will need to record the result.

As you record your results, it is likely you will start seeing a pattern of imbalance. A few signs may mean very little or nothing at all. Again, as with the risks for autism, no one measurement alone means anything. Ultimately, a true imbalance is based on the culmination of many different data points. Remember, no one has a perfectly balanced brain or body. It is the degree of imbalance that matters, and it can vary from person to person. This is why your basis of comparison is yourself: one side of your body versus the other. Normal is when both sides are the same on each checkpoint.

You do not have to do all these assessments at one time or even in any particular order. The way you feel on any given day can make a difference, so you might want to do some or all of them more than once before you come to a conclusion. We're going to start at the top and work our way down the body.

Dominance Profile

We generally think of dominance as the hand with which we write. Virtually no one can write equally well with both hands. We're either a lefty or a righty. Most people, however, will favor one hand for most tasks, though they will use the other for certain tasks. We call this mixed laterality. It is rare for someone to be all one and not the other. The degree and side of handedness, to a large degree, is genetic. For most operations, the left hemisphere controls the right side of the body and vice versa. For most people, fine motor control of rapid, sequential, and voluntary conscious movements is a left-brain operation, even in people who are left-handed. The left hemisphere is referred to as the "dominant hemisphere" because it is most often dominant for speech and verbal language, as well as handedness.

Approximately 95 percent of right-handed individuals process speech primarily in the brain's left hemisphere. More than half of left-handers process speech in their left hemisphere, just like right-handers. However, about one-fourth of left-handers process speech equally in both hemispheres.

Even though we think in terms of left or right, there are actually four main styles of handedness. Right-handedness is most common.

Approximately 70 to 90 percent of the world population is right-handed. The majority of the rest are left-handed.

Mixed dominance, also called cross dominance, is being able to do different tasks better with different hands. For example, a mixed-handed person might write better with his left hand, but throw a ball more efficiently with his right. Some tools force mixed handedness. For example, scissors are generally manufacturered for righties.

Ambidexterity, the ability to perform equally well with both hands, is exceptionally rare, though it can be learned. Even those who learn it, however, will still sway toward their originally dominant hand. Ambidexterity is almost always an unnatural condition. It has been associated with ADHD, autism, and other neurobehavioral disorders. In general, the lack of or the delay in the emergence of hand or language dominance signifies a delay in the maturity of the brain and the lack of specialization and differentiation of the two hemispheres. Hand dominance normally should be clear around age two to three.

If you are left-handed, it is a good possibility it is not the way you were genetically programmed to be. How to know? First, look at your family history. If everyone in your immediate family is clearly right-hand dominant, then chances are you should be too. Another way to test if you should be right-hand or left-hand dominant is to look at the rest of your dominance profile—your foot, eye, and ear. Dominance should be consistent, meaning you should favor one side of the body for everything. The vast majority of people should be all right-sided or all left-sided. If you have a mixed-dominance profile, it usually signifies that your brain did not fully mature and lateralize the way it was supposed to. If mixed dominance is present in you, it is possible that you could pass that on to your child. One of the best ways to know if you may have started life with a brain imbalance is to find out if you have a consistent dominance profile. The way to find out is simple.

Handedness

The best way to test for handedness is to use multiple targets. Notice which hand you use to perform a variety of common functions:

L R

☐ ☐ **Combing your hair**

☐ ☐ **Throwing a ball**

☐ ☐ **Latching the door**

☐ ☐ **Carrying a handbag or briefcase**

☐ ☐ **Zipping your jacket**

Knocking Test

This test should confirm the certainty of your observations:

> *Go to your computer and continuously strike the "f" key for thirty seconds with your left hand. Then continuously strike the "j" key for thirty seconds with your right hand. Repeat the test, first left, then right, two more times. Count the characters you knocked for each letter on each occasion. The one with the most knocks is your dominance hand.*
>
> **My hand dominance is L _____ R _____**

Leg Dominance

Test for leg dominance in the same way you did for hand dominance, by observing which leg you use for several tasks, including:

L R

☐ ☐ Kicking a ball

☐ ☐ Taking the first step on a ladder

☐ ☐ Regaining your balance

☐ ☐ Taking a step forward

☐ ☐ Lifting your leg to jump over a fence

My leg dominance is L _____ R _____

Eye Dominance

If visual information has to travel longer than necessary, the probability of making an error is increased. Reading and writing is a sequential task connected to the left hemisphere. It is from the right eye that information arrives quickest to the relevant regions of the brain. Therefore, left-eye dominance can contribute to the development of dyslexia. It does not in itself cause dyslexia, but as an element of disorderliness, it increases its probability.

Eye dominance is usually tested by a peeking task. You can tell a person's eye dominance by watching the way they use a telescope at a scenic viewpoint. Notice which eye you use to:

L R

☐ ☐ Wink

☐ ☐ Look through a peephole

☐ ☐ Look into the viewfinder of a camera, telescope, or kaleidoscope

Also take this test:

> *Stretch either arm straight out so it is level with your eye.*
> *Make a loose fist and lift your index finger vertically.*
> *Align your finger with a distant vertical straight line,*
> *such as a doorpost or the edge of a cupboard. Close one*
> *of your eyes. Open it and close your other eye. Repeat this*
> *three more times. You will notice that when you close an*
> *eye, your finger seemingly jumps sideways. The eye with*
> *which you notice the smallest or no sideways jump is*
> *your dominant eye because it is the one you used to align*
> *your finger when both eyes were open.*

My eye dominance is L _____ R _____

Ear Dominance

Ear dominance is tested by listening. Which ear would you put against a door to eavesdrop on conversation? The ear you use to answer the phone is not necessarily a good gauge because it is often ruled by hand dominance. You might answer with your left ear because you use your right hand to write something down.

My ear dominance is L _____ R _____

Head Tilt

When a person has a noticeable head tilt it is almost always a result or indication of an imbalance in the tone of your eye muscles. This imbalance in eye muscles pulls or rotates the eyes one way or the other. To keep our eyes level to the horizon, we tilt and rotate our heads to bring our eyes to the middle. The most common reason for both a head tilt and eye muscle imbalance is an imbalance in the brain and its control over these muscles and the balance of muscle tone.

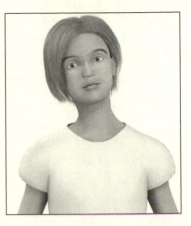

When kids show up at our Brain Balance Achievement Centers, one of the most noticeable signs we see of a brain imbalance is a head tilt. We usually tilt our head toward the side of hemispheric or brain weakness. A head tilt can be ever so subtle. If it is not noticeable in you or your partner, check the level of your ears at the earlobe. One will look lower than the other. The low side indicates the weak side of the brain.

I do ____ do not ____ have a head tilt

My head tilt is L ____ R ____

MEASURING MUSCLE TONE AND BALANCE

A brain imbalance often causes an imbalance in muscle tone between the same muscles on both sides of the body. Though the left brain controls voluntary use and dexterity of the right side of the body and vice versa, each side of the brain controls involuntary control of muscles on its own side. This means decreased brain activity in one side of the brain will manifest as a muscle imbalance or weakness on the same side of the body. The imbalance not only occurs in lateral left-right muscles, but from front to back. For example, the tone in the muscles on the back of the shoulder and arm and the front of the hip and leg on the right side will improve as activity in the right side of the brain goes up.

These very specific muscle imbalances alter body posture, facial expressions, and the way we carry and move our arms and legs. It is what I was observing in the face of the television news reporter. These differences are a very accurate and easy way to judge if you have a brain imbalance and on which side the weakness resides.

So let's take a look at you and your partner. There are two ways to go about these assessments. Look in the mirror in your natural state. That means no makeup and no posing. Or find a recent picture or video of yourself in which you were not posing. We are going to start with the head and work our way down the body, starting with the eyes.

Eyes

Take a good look at your eyes in the mirror from your brow to your lower lashes. See if you can observe a slight difference in size. This can be found by looking at where your eyelid naturally falls in comparison to the lower lash line. It is normal for there to be a slight difference in the size. A noticeable or significant difference is not normal. The larger eye is usually on the side of brain weakness. It is caused by droopy facial muscles on the side of the imbalance.

> **There is _____ is not _____ a noticeable difference in the size of my eyes**
>
> **My eye is larger on the L _____ R _____**

Eye Crossing

A weakness in an eye muscle can lead to strabismus, or what is known as lazy eye, a condition in which the two eyes are not equally aligned when focusing on a point in space. This can suppress normal binocular vision, which has an adverse effect on depth perception. Essentially, you are seeing with one eye. We see this frequently in children who come to our Brain Balance Achievement Centers. The condition can aggravate a brain imbalance because the one eye that is working well sends more signals to the opposite side of the brain.

Crossed eyes are something that frequently starts from birth, though it can develop over time. The weakness in the eye can be rather subtle and may not be that easy to detect. A trained person should be able to see a difference during an eye examination. Looking at a photograph of yourself is another way to detect it.

You should also do this test: Take your index finger and stick it out in front of you with your arm fully extended. Focus with both eyes on the finger. Then slowly start to bring the finger toward your nose. Pay attention to the point at which you see two images of your finger. This is the point at which your two eyes are not working together. If both eyes are working equally, you should be able to bring your finger within an inch or two of your nose without seeing double. I have seen people who see double at twelve to eighteen inches in front of their face. If you do not see double at all, it means your dominant or stronger eye is taking over. So, try it again and make sure to use both eyes when watching your finger.

You won't be able to tell which side is weak, but you will know if you have an imbalance in your eye muscles. If this is the case, you can follow up with an ophthalmologist, optometrist, or functional neurologist who can pinpoint the side of your weak muscles.

I start to see double _____ inches from my nose

Nasolabial Fold Depth

As we age we start forming a fold on both sides of the face from the nostril to the corner of the mouth. This is called the nasolabial fold and it gets deeper with age. Good facial muscles are taut. As we age, muscles begin to lose their elasticity, which causes the skin over them to start to sag. This is why you don't see this fold in chil-

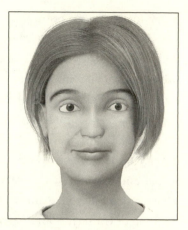

dren. In adults, the fold is naturally more pronounced in some than it is in others.

You are not examining your face for the depth of the fold. Rather, you are looking for a difference in the depth from one side to the other. Also, look at the wrinkles and creases all over your face. The side with the least impressions is the side of brain weakness. A good way to understand this is to think about Botox. Botox is a neurotoxin that paralyzes or weakens muscles, which reduces the appearance of wrinkles. If you have weaker facial muscles on one side of your face, you will have fewer wrinkles on that side. This is also usually the side of your brain that is weak.

The nasolabial fold is less pronounced on the L ＿＿ R ＿＿

Mouth Droop

Look in the mirror and observe the corners of your mouth in a relaxed position. Check to see if one side is lower than the other. You should also be able to observe this as you speak. It can also become less or more noticeable when you smile naturally.

Most people tend to favor the left side of the face when expressing emotion. This is because it is easier for the right hemisphere of someone facing you to see the left side of your face. The right brain is dominant for reading and expressing emotion.

Both corners of the mouth rise to smile. However, in a natural neutral position, one side can be lower than the other. The side that is lower is the side of brain weakness.

I do ____ do not ____ have a mouth droop

My mouth droops more on the L ____ R ____

Soft Palate

The soft palate muscles at the back of the roof of the mouth help us swallow and produce sound. When we speak or make a sound, the soft palate goes up.

For this test you will need a flashlight or a penlight. Look in the mirror or in your partner's mouth and aim the light toward the back of the top of your throat. You will see a little white line through the middle that attaches the right and left sides of the soft palate muscles. Say "ahhhh" loudly and repeatedly. You will see the roof of your mouth go up. What you are looking for is to see if one side goes up faster and/or farther than the other. You may also notice that the white stripe in the middle is pulled to one side more than the other. The side that goes up more slowly, or the side that the white stripe moves away from, indicates the side of brain weakness.

My soft palate appears normal ____ does not appear normal ____

The soft palette is weaker on the L ____ R ____

Tongue

Like the soft palate, the tongue is really two muscles joined in the middle. When you stick your tongue out, the two muscles push hard together as if they are equal. If one side is stronger it will push harder against the other side, making it deviate to the other side.

Look in the mirror and stick out your tongue. The side to which it deviates is the side of tongue weakness. The side of tongue weakness is usually the side of brain weakness.

My tongue muscles appear normal _____ do not appear normal _____

My tongue deviates to the L _____ R _____

MEASURING POSTURE

When the brain is out of balance, it affects the balance of muscles from side to side as we are standing still. It will also affect coordinated balance and tone of the muscles that make walking and other fluid motion possible. It also can cause an imbalance in tone between the front and back muscles in the extremities. These are the muscles in the shoulders, arms, and front of the hip and leg. These muscles and the brain work in tandem to make movement possible.

Our postural muscles help us capture and harness gravity and send it to the brain to stimulate growth and coordination. Gravity is by far the most important stimulus because it is constant. It helps to form the baseline idling speed of the brain. Because this stimulus is so important, the brain will increase the tone of the muscles that fight off and essentially "capture" gravitational pressure and send it to the brain. These muscles have several unique features, including the brain's ability to change the sensitivity of muscle receptors to increase or decrease the amount of stimulus the muscles send to the brain at any given moment. This is really what muscle tone is all about. It is the amount of sensitivity in the muscle receptors and the tightness or looseness of the muscle. The more active the brain is, the better your muscle tone will be.

Elbow Bend

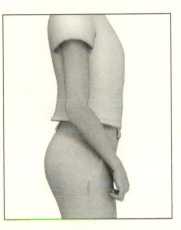

When the brain is weak, the muscles in the back of the shoulder, arm, and hand become weak in relation to the front of the shoulder, arm, and hand. This causes the two arms to appear different when bent at the elbow in a relaxed position. There is no way to test for this. You can only do so through observation, so take a close look at the illustration where the elbow is more bent and the arm is more flexed. This causes the arm to rotate in so you see more of the back of the hand. The abnormality appears on the same side as brain weakness.

> **My elbows appear to be the same _____**
> **not to be the same _____**
>
> **The elbow that is bent more is the L _____ R _____**

Hand Bend

Another clue that can only be made through observation is hand bend, where the hand will be more closed on the same side where the elbow is more bent. The fingers will look more curled compared to the other side.

Hand bend is more noticeable on the L ____ R ____

Shoulder Tilt

We all have uneven shoulders, and it is usually noticeable when you look at yourself standing naturally in front of the mirror. Your shoulder should be lower on the side of your dominant hand. If it is lower on the opposite side, it indicates a weakness in the brain on the side where the shoulder is high. However, if you have a short leg, your shoulder will be lower on that side. You can check for a physically shorter leg with a tape measure by measuring from your hip to your anklebone. If you still can't tell, go with the result of the head tilt test.

My shoulder is lower on the L ____ R ____

Standing Position

Normally when you are looking at someone standing straight and tall, the arms are rotated inward and hands are by the side, with the back of the thumbs aimed straight ahead. However, if the shoulder is rotated forward, the back of the hand, rather than the thumb, will face forward. This means that it is not just the shoulder that is out of balance but the arm and hand too.

Check your own stance when you are unconsciously standing at attention.

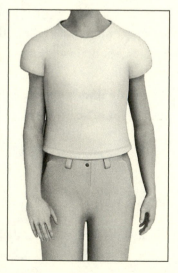

If it is not as I just described, it is a sign of a brain imbalance. The imbalance occurs in the side of the brain of the out-of-position arm.

This type of imbalance is most often responsible for problems such as rotator cuff tendinitis, tennis elbow, and carpal tunnel syndrome and it usually occurs on the limb with the weak muscles.

When standing at attention, my arms are perfectly aligned _____ are not perfectly aligned _____

My arm is out of alignment on the L _____ R _____

Knee Bend

The same relationship just described in the shoulders, arms, and hands also apply to the hip, leg, and foot, but it is reversed. When the brain is weaker on one side, the muscles in the front of the leg get weaker and the muscles in the back of the leg, such as the hamstring muscles, get stronger or have more tone. These are the muscles we need to stand upright and tall. When these muscles get out of balance it can happen as a result of a brain imbalance on the same side because the less active or more immature side is not controlling the muscle tone optimally. In this case the leg will start to bend at the knee as the muscles in the back start to overpower the muscles in the front. A leg that bends more at the knee with the foot turned slightly inward is a sign of a brain imbalance on the same side.

Another way to see this is by looking at the person while seated. A leg that is slightly bent will cause a slight bend forward at the waist, which takes away from your ability to comfortably sit and stand up straight. People with this weakness can get patella tendinitis or plantar fasciitis on the side of weakness. Back pain is also common. If you have these symptoms, a professional can confirm the diagnosis.

Measuring Muscle Strength

Another way to verify these findings is to check your muscle strength on one side of the body compared to the other. It is normal for your dominant hand to be slightly stronger, but not significantly stronger. If your dominant hand is weaker while doing these tests, then you might have a brain weakness on the same side as the weakness. An exception to this rule is a weakness caused by injury to the muscles of the arm, hand, or leg.

Thumb

Make a fist with both hands and give the two-thumbs-up signal. Have someone apply pressure with the palms of their hands to both thumbs at the same time, attempting to push them down. If one side is significantly weaker, the thumb on that side will go down easier than the other thumb. This is a sign of a brain imbalance on the same side of the brain as the thumb weakness.

My thumb is stronger _____ weaker _____ on my
R _____ L _____ hand

Big Toe

The big toe will be the same. Lie on your back with your toe pointing straight up. Extend your big toes back up toward your head and have someone apply pressure as they did with the thumb, attempting to push them back down. A significant difference, meaning one side is noticeably weaker than the other when pressed, indicates a brain imbalance on the same side as the weakness.

My big toe is stronger _____ weaker _____ on my
R _____ L _____ foot

Hand Grip

Grip strength works opposite. When you consciously close your right fist, it is being controlled by your left brain. The dominant hand should only be about 10 percent stronger.

To test hand grip have someone place their fingers or thumbs in both of your hands. Grip them tightly and ask the person to try to pull their fingers free in both hands at the same time. Repeat two more times. If one side slides out easier than the other this may signify a difference in muscle strength. The weaker side will repeatedly not be able to hold on to the fingers because your grip is consistently weaker on one side. This is usually opposite of the side of the weaker hemisphere. Another way to accurately measure this is by using a hand dynamometer, a device that measures force. A functional neurologist can perform this for you.

My hand grip is weaker on my L ____ R ____

Testing Your Sensory System

Along with our muscles, our senses play an important role in brain stimulation. To this end, you will also be looking for imbalances in your sense perception.

Sound

You can compare your hearing by simply listening and judging if sound seems louder in one ear than the other. Here is a test using an iPod:

Turn some music on at low volume so you can barely hear it at close range. Put the device in your right hand and extend your arm straight out so the iPod is facing your right ear. Slowly bring it closer to your ear. Note the distance at which you can first detect the sound. Do the same with your left ear. This test is even more precise if you have someone else moving it for you. Close your eyes as it is being performed.

Another way to do this test is to use one earpiece of your headset. Turn the volume of your iPod all the way down and slowly turn up the volume until you just start to detect sound. Take note of the number or level of the volume. Do the same with the other side and compare if one ear is weaker.

The side where you can detect sound closest to the ear or the side on which the volume is louder would be opposite the side of brain weakness. However, it isn't necessarily a sign of a brain imbalance because there are many reasons why your hearing could be better in one ear. The number one reason would be earwax. If you detect a difference, have it checked out by your doctor.

I detect no difference in sound ____

Sound appears to be stronger in my L ____ **R** ____ **ear**

Touch

Close your eyes and have someone lightly touch one arm and then the other with identical light pressure. Next, have the person touch both arms at the same time. Note if the feeling is different on each side. Do the same on the legs. The side in which the sensation is the lightest is an indication of a brain weakness on the opposite side, especially if both the arm and leg are less sensitive on the same side.

I detect no difference in touch sensation ____

Touch appears to be more noticeable on my L ____ **R** ____ **side**

Smell

Smell is a very important sensation, as it is often the first sign that something is wrong in the brain. For example, losing your sense of smell is an early warning sign of Alzheimer's disease. It is also associated with schizophrenia and we have seen that it is almost always abnormal in autism.

Here is a simple test. Fill a small vial with coffee, pinch one nostril closed, take a deep breath, and gradually bring the vial toward your nose. Note the distance at which you can detect the smell. Repeat with the other nostril. Also note if there is a side in which the smell is strongest.

You should also do this with your eyes closed. Have someone bring the scent toward you in the same way. If you notice a difference, try it with different smells to see if you can identify them equally.

Smell is the only sense that is controlled by the same side of the brain, so the side with the weakest sense of detection is the same side as a brain weakness.

I detect no difference in smell sensation _____

My sense of smell is weaker in my L _____ **R** _____ **nostril**

Temperature Sensation

A good way to test temperature sensitivity is with four metal spoons. Put two in the refrigerator to cool them and the other two in hot water to warm them. Close your eyes and have someone touch both your arms with the cold spoons. Note if the sensation feels the same on both sides. If one feels weaker, note which one. Do the same with your legs. Repeat the same experiment with the warm spoons. The side where the temperature sensation is weakest, especially if it occurs in both your arm and leg, is opposite the side of brain weakness.

I do _____ do not _____ detect a temperature difference

Temperature feels stronger on my L _____ R _____ side

Balance

Sense of balance is coordinated by the inner ear, which is called the vestibular system, and involves the brain, particularly the area at the base called the cerebellum, and your muscles. If you've ever had a little too much to drink, you know what happens to the vestibular system. It's the reason it's referred to as being tipsy.

If you are prone to bumping into furniture or tripping over things, it's a sign that your sense of balance may be off. Here's a way to check: Stand relaxed with your hands by your side. Bend your right leg back at the knee and hold for as long as possible. Do not grab on to your leg and do not hold on to anything for balance. However, stand close enough to something you can grab on to when you begin to lose your balance. Note how long you could hold the pose. As you start to lose your balance, note the direction in which you start to fall. Do the same with your left leg. The side to which you lean is usually the opposite side of your brain weakness.

Here's another test: Stand facing forward and extend your arms straight out and clasp your hands together with your fists in a thumbs-up position. Keep your eyes on your thumbs and slowly turn your head to the right and bring your head back to center. Do this ten times in one direction or until it makes you feel dizzy. After the dizziness passes, do the same to the left. Dizziness occurs on the side opposite of brain weakness. Dizziness, which is also referred to as vertigo, is almost always a result of an imbalance in the inner ear or brain.

This exercise does ____ does not ____ make me dizzy

I felt dizzy turning my head L ____ R ____

If you do not get dizzy, then compare the difficulty or ease of keeping your eyes straight ahead while turning your head. The side on which focusing your eyes is more difficult is opposite the side of brain weakness.

Adding It All Up

Add up all the areas in which you did not detect an imbalance and the areas in which the imbalance signified the left brain and right brain. What is significant varies for each person, but anything more than a 10 percent difference between the two sides is probably too much and may represent an imbalance. Keep in mind that how you score depends on how you are functioning on the particular day in which you perform the test. So if your score is greater than 10 percent—or if it is less than 10 percent, but you don't trust the results—you can repeat these exercises on another day. It is not necessary to repeat the exercises in which no imbalance was indicated.

Since getting to balance is rather subjective, you can make it more objective by quantifying the results. You can do this by establishing a number system from 0 to 3 in which 0 represents no imbalance, 1 represents a mild imbalance, 2 represents moderate imbalance, and 3 represents a severe imbalance. Your goal is to be in the 0 to 1 range.

You should also consider these results in comparison to the other tests you've taken in this book and risks you've noted. The more information and the more data points you have, the more accurate your self-assessment.

CHAPTER EIGHT

Your Lifestyle and the Stress Response

Stress. There's no escape. It's nearly impossible to get through a day without thinking it, feeling it, or hearing about it. Yet ask one hundred people what stress is and you'll likely get one hundred different answers. There's a reason why.

Stress is hard to define because it means something different to different people. Take rock climbing, for example. To me, just the idea of climbing the sheer face of a mountain with no support ropes is frighteningly stressful. To the next guy, it's his version of spending the afternoon at the spa. It's the classic good stress/bad stress scenario: exhilaration (good) versus fear (bad). However, if you were to measure the vital signs in cowardly me and the person without fear, you'd find our metabolic reaction would be similar, if not the same. The prospect of the climb would make our heart race, our blood pressure would go up, and we'd start breathing faster. Our stress response would be on full alert—mine out of dread, the other guy's out of anticipation.

As you've probably read in magazines or heard from your doctor, a certain amount of stress is good for the body. However, it is the more fleeting type, like the example I just described. You're making a major presentation to your board of directors. You're getting ready to leave on a big trip. Your in-laws are coming for the week. Short spurts of stress, either positive or negative, generally aren't going to harm you. The kind that is bad for you is the kind that hangs around—the loss of a job without any prospects of another, dealing with a chronic disease or pain, an abusive marriage, or being a single parent with a job and many mouths to feed.

Stresses in life increase our stress *response*, a physiological reaction that we cannot feel but is always with us. When our stress response is too high for too long it wears on our bodies and our brains. I believe a heightened stress response in both parents, but especially the mother, may turn out to be the single biggest risk factor leading to autism. I believe it plays a major role in turning on or turning off the genes that your children can inherit that can lead to autism. I also believe it is the most important risk factor prospective parents need to monitor if they are planning to have a child.

The Stress Response

Most of us think of stress as a damaging consequence of some negative or adverse event or situation. However, this is not how stress works or how it damages our health and affects our brain. Stress and the stress response are two different things. Even the coolest dudes who claim they don't know what stress is have a stress response. It roams through our bodies 24/7. This is rarely the kind of stress people think about when they talk about bad stress, but it should be. For many of us, it is more often working against us than for us. It's more damaging than the type of stressful fright you get when you're three inches away from a head-on collison or the stressful thrill from taking a seventy-mile-an-hour plunge on the Wicked Twister roller coaster. In fact, you don't have to feel any kind of stress at all to get a negative reaction to the stress response. Let me explain.

The stress response is operated by two subdivisions of the nervous system: the sympathetic nervous system and the parasympathetic nervous system. They basically work in opposition to each other and create opposite physiologic effects on our bodies. They also jump into action at different points in time.

The sympathetic system is also referred to as the fight-or-flight system. It jumps into action even before we know what is happening. It's what makes us swerve to avoid a collison or jump back on the curb when a bus comes out of nowhere. It activates a physiological response to danger. The heart races, the blood vessels to our muscles and pupils dilate, our digestive system shuts down, our organs constrict, the liver quickly breaks glycogen down to glucose to give our muscles energy to burn, and we breathe rapidly to take in more oxygen to keep our muscles working. Many of our other metabolic functions turn off as well, such as our detoxification system, making us more vulnerable to the effects of environmental toxins. We perspire to keep cool, the immune system goes in repair mode creating inflammation, and the brain is aroused and vigilant. All of this happens in an instant to prepare the body for rapid and decisive action. This is our body's way of responding to a hostile environment. It's our physiological survival mode. It's why we can exhibit unworldly strength in times of crisis. This is the primary focus of this part of our nervous system. It is automatic. Do nothing and it will work on its own.

The sympathetic nervous system is so essential to our survival we are born with it fully intact. Survival is the main drive of a newborn. Our basic survival needs—breathing, a beating heart, movement, and the ability to cry—are all a baby initially needs to survive. These activities are all subconscious reflexes that work automatically.

Because the sympathetic nervous system is so spontaneous and swift, we need to keep it from going out of control. This is where the parasympathetic nervous system comes in. It gets us back to "rest and digest" mode. Once the danger is gone, we need to be able to shut the sympathetic system down as quickly as possible.

The sympathetic system is the red alert in our bodies. So, to a certain degree, it is always on the ready, meaning it has an "idling speed," much like an automobile. Before you turn on the ignition in your car, you have to step on the brake to prevent the car's natural ability to lurch forward. The sympathetic nervous system is like the gas pedal and the parasympathetic nervous system is the brake. Take your foot off the brake, and the car will start to move. To stop it from moving, we step on the brake. However, as you sit there with the engine running and your foot on the brake, the car idles.

As any mechanic will tell you, idling speed is important for the health of the car. If it idles too fast, it will burn up gas unnecessarily and wear on the engine. Eventually, you'll start to have engine trouble. If the idle is too slow, it will stall and the car won't move. You want the idling speed to be as low as it can go to keep the engine purring, but not racing. The same goes for your body.

The activity and balance of the sympathetic and parasympathetic nervous systems set the idling speed of the body. However, there is a second idling speed—the one in the brain. The brain's idling speed is essentially the rate at which it can process and share information. As the brain grows and develops, its idling speed gets stronger and faster. As the idling speed of the brain grows faster, the idling speed of the body or the sympathetic stress response goes down, putting less strain on our body. This is very important, because it is a mechanism that helps extend our long-term survivability.

So ideally, you want a very fast, coordinated, synchronized, and balanced idling speed in the brain, which gives you a relatively slow idling speed of the sympathetic nervous system or stress response.

This is how we define optimal health. When the idling speed of the body increases or when the sympathetic nervous system is too active, it means we have a high stress response. That's not good.

As a child's brain matures and becomes more active, coordinated, and synchronized, we can actually see that as the baseline activity in the brain goes up, the baseline activity of the stress response goes down. As we become less physically and mentally active as adults, our brain's idling speed goes down, which increases the stress response.

A high stress response is hard on the body because it floods the body with stress hormones, most notably cortisol. It's how we end up with high blood pressure and a faster heart rate. It's why we need to catch our breath when we're out of shape. Science is coming more and more to the realization that the stress response is a key player initiating chronic disease.

The idea that the brain is involved in chronic illness is counterintuitive. When we think of keeping the brain active, we think in terms of mental activity—reading, learning, or doing puzzles. We assume that if we are mentally active, then our brain must also be active, and therefore we would assume that the baseline idling or processing speed of our brain would be active as well. However, the brain needs *physical* activity. Movement of our large muscles is one of the biggest sources of stimulation for the growing brain. It is why kids need to be outside playing ball and climbing trees instead of sitting for hours on end staring at the television or computer screen. Likewise, physical activity is important in setting the idling speed of the brain in adults as well, and by extension the stress response.

So, as you can see, an increased stress response does not necessarily have anything to do with an increase in the stressful siutations of life. The reason this is so important is because high levels of cortisol that signal a high stress response are common in children with autism, as well as their parents. Many studies have found that children with autism have abnormally high levels of cortisol in their bodies. A high stress response means the body is churning out high levels of cortisol. And here's the clincher: Cortisol is a steroid hormone involved in gene expression.

What Happens in the Womb Lasts a Lifetime

The fetus is very sensitive to the high cortisol levels of a mother with a high stress response. The medical world unfortunately found this out the hard way after seeing the long-term effect from treating fetuses at risk for premature birth with a synthetic form of cortisol, which helps promote lung development. Underdeveloped lungs are a life-threatening risk factor in premature birth. Doctors and scientists started to notice over time that people who had received this treatment in the womb had a lifelong heightened stress response and a greater incidence of heart disease, diabetes, and other chronic illnesses. They also found these treatments predisposed children to behavioral problems, including anxiety disorders, depression, schizophrenia, a propensity toward substance abuse, ADHD, and even autism. The reason? Synthetic cortisol treatment essentially mimics maternal stress.

When a pregnant women is mentally stressed or she has an increased stress response, her production of cortisol goes up, which is transmitted to the fetus through the placenta. Scientists believe this may permanently readjust the settings of the stress response in the fetus in a way that may increase sensitivity or hyperresponsiveness to stressful events in life. It possibly also raises the baseline idling speed of the stress response.

Any number of things can induce a chronic stress response in a mother-to-be, from an actual or perceived stressor, such as a bad or abusive marriage or a high-pressure job, to exposure to harmful environmental chemicals or illness, to social isolation or anxiety, to chronic conditions such as depression, high blood pressure, and diabetes. War and natural disasters are two other reasons that will increase the stress response in a mother-to-be. Studies show that children of mothers who suffer from post-traumatic stress disorder (PTSD) during pregnancy are more likely to experience the same disorder. For example, records show that children born to mothers who lived through the Holocaust were more prone to develop PTSD, even though they never experienced the Holocaust. Records also show that all children of Holocaust survivors were more prone to depression. Children whose fathers survived the Holocaust, however, were not affected by their trauma. It is well documented that PTSD is related to increased right-brain activity and decreased left-brain activity. Although this trait is the opposite of how autism develops, it still can act as a blueprint for how autism can be passed from mother to fetus.

Research is beginning to see similar signs in the children of women who were present and pregnant at the World Trade Center on September 11, 2001. As would be expected, a number of them ended up with PTSD. They also gave birth to babies with an elevated stress response. These children, most likely, will also be more susceptible to anxiety, depression, and even PTSD as they get older.

Males are more affected by maternal stress, and it is possible this is the reason why boys are more prone to autism than girls. Studies found that when pregnant guinea pigs were treated with synthetic stress hormones, only their male offspring were born with an elevated stress response. Research indicates that this has a lot to do with the role testosterone plays in the womb. Development of the male form is dependent on elevated testosterone. However, the female form is not.

How Disaster Influences Genes

One of the best-known examples of how stress can affect future generations took place in the western Netherlands around the end of World War II. Toward the end of the war there was a scarcity of food in the region, which continued to get more severe. Food rations were distributed to women at 1,000 calories a day, less than half of the 2,300 calories that was considered to be a normal intake. As matters got worse, the rations dropped to 600 calories, most of which came from bread or other carbohydrates. Over the course of almost a year, 22,000 people in the western Netherlands died, most from starvation.

Soon after the war, it became apparent that a natural experiment was taking place. People in the western Netherlands had been exposed to prolonged severe stress but the other half of the country had not. This became known as the Dutch Famine Birth Cohort Study. Through obstetrical records, researchers found that many babies born during the famine weighed significantly less than babies born before the famine—no surprise there. The long-term effects of the famine didn't start to become apparent until the 1960s when children who were in their mother's womb during the famine turned age eighteen. They found that fetuses exposed to the famine during the second and third trimesters showed significant levels of obesity, roughly double the levels of those born after the famine. This was in spite of the fact that people exposed to the famine during the third trimester were abnormally small at birth. They also found that people exposed to the famine starting in the first trimester were born larger than average, suggesting that the mothers' bodies initially tried to compensate for lack of food in some way, perhaps increasing the size of the placenta.

Other subsequent studies looked at psychological outcomes. Researchers found a significant increase in the risk of schizophrenia and emotional disorders, such as anxiety and depression, and antisocial personality disorder, particularly among males. Back then, there was virtually no awareness of autism and Asperger's, but I would bet if those studies were conducted today, the emotional traits they chronicaled would be diagnosed as autism spectrum disorders.

As the study population started hitting their fifties during the 1990s, the researchers found that people prenatally exposed to the famine were more prone to obesity than similar people who were not exposed to the famine in the womb. They also had a higher incidence of high blood pressure, coronary heart disease, and type 2 diabetes. Follow-up studies later in the decade found this trend continued to get worse.

Curiously, the researchers ascertained that the famine itself was not the source of a lifetime of health problems, it was the prenatal time of exposure that made the difference. Women exposed during the first trimester had an increased risk of breast cancer. Those who were exposed during the second trimester had lung and kidney problems. Altered glucose tolerance and diabetes were seen more in those exposed in the third trimester.

Now let's fast-forward to disasters of more recent times—hurricanes. Studies looking at hurricanes that hit Louisiana between 1980 and 1995 found a direct relationship between an increase in the number of children who were subsequently diagnosed with autism and the severity of the storm. The more severe the storm, the greater the number of children with autism. This was especially true for the offspring of women who experienced the storms during their second and third trimesters.

All of these stories are great examples of an epigenetic effect in which the environment coupled with a high stress response can influence the expression of brain-building genes.

The environment affects genetic expression in an indirect way through the genes that reside in our cells. Different kinds of cells respond differently to the same environmental factor. The brain cells will react one way to famine and the liver cells or pancreas cells will react another way. This is why something like the Dutch famine could affect so many different types of cells. If we looked at the individuals who went through the famine versus those who did not, we would see different activity of different genes in the different tissues, such as the brain and the liver. This altered pattern of gene expression is what epigenetics is all about.

What happened was that the children of mothers who had a high stress response tended to become overweight or obese later in life and have all the health problem that go with it. They were also more prone to psychological and neurodevelopmental disorders. This is the trend we are seeing in society in general. There is an epidemic rise of not only autism and other neurobehavioral disorders but also of obesity, diabetes, heart disease, and autoimmune disorders. They all evolve in the same vicious cycle. I do not believe this is coincidental. I believe the one factor that can explain all of these increases is a general increase in the idling speed of our stress response over the past two decades.

This response may in part be due to more stressful living, but I believe it has much more to do with our lifestyle, most notably sedentary behavior. Reduced physical activity is decreasing the idling speed of our brain and slowly increasing the idling speed of our stress response. We think that we keep our brain active by putting it through its own kind of "physical" exercise, like doing crossword puzzles or learning something new or playing computer games. But in fact, the stimulation that best builds the brain in babies and toddlers is movement of the big muscles. It's becoming more and more obvious that this type of big-muscle movement is also what keeps the adult brain healthy and its idling speed low. This is why I am also stressing that routine exercise is essential for anyone who wants to have children.

Our dependence on technology means we move less and less, and this is having a major impact on our brains, and the bodies and brains of our children. Based on what we have seen from the studies of the

Dutch famine, if more mothers have a higher stress response and increased cortisol, we would expect to see more and more children with obesity. And that is exactly what is happening.

WHY ARE WE SO STRESSED?

High cortisol levels, inflammatory chemicals in the blood, high blood pressure, and a high heart rate are all indicators of a high stress response. Studies show that these physiological effects are on average significantly higher in the United States than they were twenty years ago. If we could go back and measure the stress response of people fifty or one hundred years ago, we would surely find that these same physical measurements of the stress response would be even lower. The natural response to this is often, *Well, that makes sense because people back then didn't have the stress we have today.* But let's think about that for a moment.

We say life today is more stressful because of its hectic pace. In reality, life today is pretty cushy compared to our ancestors. Imagine living without the modern conveniences of indoor plumbing, running water, automatic dishwashers and washing machines, cell phones, and moment-to-moment news and weather warnings. Compare it to an existence where you had to work the land twelve hours a day, seven days a week, and one good storm could wipe you out or a pestilence could put an entire family on the brink of death. Imagine what life was like, even here in the United States, during the great wars, where food was rationed and the men of war were gone for years with not much word of where they were or if they were safe. To me, that was a truly stressful life. Do we really think that we are more stressed today than our pioneers who rode on wagon trains or our ancestors who survived the Great Depression? By comparison I don't think getting stressed out over having to catch the commuter train to a high-pressure job in which you sit in front of a computer and are on the phone all day comes close to the kinds of stress our forefathers had to deal with.

When you look at what killed our ancestors, we see more death from infectious diseases and famines than from the stress-related chronic diseases that are killing people in modern times. There is a new label attached to today's leading killers—cardiovascular disease, stroke, diabetes, and many forms of cancer. They are called lifestyle diseases. The true source of modern-day stress isn't stress itself, but a ratcheting up of the stress response. When you look at people with these conditions you can see they have a chronically high stress response, and along with it, chronic inflammation.

How Do You Know?

The most direct way to find out if you have chronically high stress is through a saliva test that you can perform at home on your own. I explain all the various ways you can test your stress response in Chapter 10. However, if you listen to your body, you can detect other clues, because the high cortisol levels that go along with the high idling speed of the stress response have multiple health implications.

Cortisol increases the flow of glucose out of your muscles and into the bloodstream in order to increase energy and physical readiness to handle the stressful situation or threat. The increase in blood sugar increases a release of insulin, which helps move sugar out of the blood and into cells for energy. This is fine in short spurts—the fight-or-flight scenario for which the stress response is intended. However, if levels of cortisol are chronically high, blood sugar levels are high and levels of insulin are chronically high. When blood sugar levels are consistently high, it increases insulin resistance, the cells become less responsive to insulin, and this keeps the sugar in your blood, putting you at increased risk for diabetes. This is especially risky during pregnancy.

Blood sugar that isn't going into cells for energy is basically converted into triglycerides and ends up as fat. Converting sugar into triglycerides takes a lot of energy, so the combination of less energy in the cells and the use of more energy to make triglycerides will cause a person who is suffering from this feel incredibly tired after eating a meal. This is why someone with low blood sugar will get a sudden burst of energy after eating.

So even without a blood test, there are signs. If you feel either a huge burst of energy after you eat or you feel like you need to take a nap or drink a cup of coffee to pep you up, then you have a problem regulating your blood sugar. In most cases, this is going to happen as a result of producing too much cortisol. It's just another sign of a stress response that is idling too high.

Knowing you blood sugar level is another important factor to be aware of because it is a good indicator of your body's stress response. This is important because the stress response not only affects blood sugar and insulin levels, it also has a direct impact on testosterone, estrogen, and progesterone levels. This is another example of how risks pile up and how they converge to increase the risk of autism. I will show you how to test for them as well.

Do you have high blood pressure or a high pulse rate? Babies are born with a high pulse rate that gradually comes down as the brain matures. If the heart rate is high it can be looked at as a more immature heart rate. It can also be a sign that the stress response is high. This is especially true in people in their childbearing years.

A high stress response increases sweat production. The skin can feel moist and clammy. You can have chronically sweaty hands and feet.

People with a high stress response can have a negative reaction to cold. Raynaud's disease is an autoimmune disease in which the extremities can get so cold they can turn white from just sticking them hands in the icebox to retrieve a few ice cubes.

Do you have allergies or asthma? There's a clear connection between stress and asthma. Studies show that stress can trigger an asthma attack or allergies just as much as smoke, pets, polluted air, and anything else. Allergies and asthma are autoimmune diseases. Autoimmune diseases create chronic inflammation. Chronic inflammation and a high stress response are interrelated. When you have one, most likely you have the other, as you are about to find out in the next chapter.

CHAPTER NINE

When Immunity Gets Out of Balance

Where there is chronic stress, there is usually inflammation, and where there is both, there most likely is a brain imbalance. Inflammatory chronic illness and autoimmune diseases are part and parcel of an immune system that is out of balance.

Food sensitivities and leaky gut syndrome, which are common traits in autism, are among the many autoimmune disorders that have witnessed a dramatic rise over the last two decades, just like autism. In fact, most of the chronic illnesses we see today are autoimmune related. According to statistics, approximately fifty million Americans, or 1 in 5 people, have some type of autoimmune disease. A majority of them, an estimated 75 percent, are women. I believe this is because women lean toward a left-brain cognitive style more so than men. This makes taming the immune system critical when a woman is considering pregnancy. As I pointed out in Chapter 4, immune problems are common in both autistic children and their mothers. I do not believe it is a coincidence when we discover that a woman with a right-brain deficit, a chronically elevated stress response, and an overactive immune system ends up with a child who has autism.

During healthy, successful pregnancies, there appears to be a natural reduction of inflammatory chemicals in the mother, which helps lower the chance of maternal rejection of the fetus and protect it from the mother's inflammatory response. Studies have found that healthy women report fewer symptoms of arthritis and other autoimmune or inflammatory disorders, even symptoms of multiple sclerosis, during pregnancy. To me this implies that increased inflammation in the mother and the child are not good during pregnancy. Research bears this out. For example:

- Numerous studies have established that stress and distress in the form of depressive-like symptoms are predictive of exaggerated levels of inflammatory markers in pregnant women.

- Elevated pro-inflammatory markers in maternal blood and amniotic fluid are causally implicated in risk of preterm delivery.

- Inflammation can contribute to preterm delivery by promoting hypertensive disorders of pregnancy. High levels of circulating inflammatory markers have been found in women with gestational hypertension and preeclampsia. Both have been found to raise the risk of autism.

- One study found that pregnant women who reported greater perceived stress had higher circulating levels of inflammatory markers and lower levels of anti-inflammatory markers in their bloodstream.

When we look at the hemispheric control of the immune response, we see that decreases in right-brain activity in the mother can result in a similar situation in the child, which would increase the inflammatory response in both mother and child. Inflammation is a risk

THYROID DYSFUNCTION:
AN UNDERDIAGNOSED DISORDER

The most common autoimmune problem, and the most underdiagnosed, involves the thyroid gland. Many women suffer from some type of thyroid problem. According to the American Association of Clinical Endocrinologists, more than twenty-seven million Americans suffer from thyroid dysfunction, half of whom are never diagnosed. Of the diagnosed cases, half are due to a disorder called Hashimoto's disease, in which the immune system attacks the thyroid gland and destroys tissue. For this reason alone, any woman contemplating pregnancy should be aware of the most common symptoms of low thyroid function:

- Fatigue
- Constipation
- Unexplained weight gain or difficulty gaining weight
- Morning headaches that wear off as the day progresses
- Depression
- Hypersensitivity to cold weather
- Poor circulation and numbness in the hands and feet
- Muscle cramps while at rest
- Increased susceptibility to colds and other viral and bacterial infections and difficulty recovering from them
- Slow wound healing
- Excessive sleeping or insomnia
- Chronic digestive problems, such as low stomach acid
- Itchy, dry skin
- Dry or brittle hair
- Swelling, especially of the face
- Heart palpitations
- Inward trembling
- Increased resting pulse rate
- Night sweats

Testing for thyroid autoimmunity is complex. If you have or suspect you have the disease, talk to your doctor or a functional neurologist about getting tested before getting pregnant. There are nutritional and other interventions that can help you overcome the disease, but they are too cumbersome to detail in this book. The best source on the subject is Dr. Kharrazian's book. In fact, I recommend that anyone with any of the risk factors discussed in this book should read it. Like myself, Dr. Kharrazian is a practitioner of functional medicine and functional neurology and understands autism and autoimmune disease from this perspective.

factor for preterm delivery, which is a leading cause of infant death. Surviving preterm infants are at high risk for serious health complications including respiratory, gastrointestinal, nervous system, and immune problems—all of which are implicated in autism. Longer-term risks include cerebral palsy, mental retardation, learning difficulties, hearing and vision problems, poor growth, and possibly ADHD and autism.

My good friend Dr. Datis Kharrazian, author of *Why Do I Still Have Thyroid Symptoms?*, is a world-renowned expert on autoimmune disease and childhood developmental disorders, particularly autism. "I am finding an autoimmune disorder is often at the root of autism, with the immune system attacking the brain or nerve tissue of the child with autism," he says. He notes that when a woman goes into pregnancy with a leaky gut, insulin resistance, multiple food intolerances, and a high stress response, "she is putting her baby at risk for developing one of the increasingly common, modern health disorders, including autism spectrum disorders, eczema, asthma, food allergies, and food intolerances."

SEARCHING FOR ANTIBODIES

It is not always possible to identify exactly what triggers a person's genes to turn on an autoimmune response. It can be caused by almost anything; however, I believe in most cases it starts with a brain imbalance creating an imbalance between the parasympathetic and sympathetic nervous systems, which then goes to the gut and ends up in the immune system. This is why leaky gut syndrome is common in people who have a brain imbalance, food sensitivities, or other autoimmune disorders.

Some autoimmune responses are easy to correct. For example, in the case of allergies to food or substances, all you need to do is remove the food from your diet or the substance from your environment. However, when the immune system starts to recognize your own body as the antagonist or the antigen, we can't just remove that body part or tissue, so it's not all that easy to fix. In medicine the answer to autoimmunity is to massively suppress the whole immune system with prednisone, a man-made form of cortisol, or some other steroid. However, this does not correct the actual problem, and it leaves you vulnerable to chronic infections. Also, as you read in Chapter 8, taking artificial cortisol can have a devastating epigenetic effect on the fetus.

The only resolution is to restore balance to the immune response at the source. This means first balancing the brain and then balancing the pro-inflammatory and anti-inflammatory pathways. An elevation of either of these systems relative to the other can result in an autoimmune reaction. It is important to test the balance of these systems. When the pro-inflammatory system is overactive and the anti-inflammatory system is underactive, as we generally see in autism, it creates a domino effect. The imbalance releases immune chemicals that antagonize the body and cause inflammation. Inflammation increases the stress response, which reduces gut function and alters hormones and blood sugar. Over time the body releases more cortisol in an unsuccessful attempt to reduce inflammation. The result is both high inflammation *and* high cortisol, a disastrous combination. The result is not one problem, but many problems. And we see all of them in autism and in many of the parents of children with autism.

LOOKING FOR AN IMMUNE IMBALANCE

If you have an autoimmune disease or a family history of autoimmune disease, or if you have allergies or food or chemical sensitivities, then you most likely have an imbalance in your immune system. If you get frequent and stubborn viral infections, such as chronic colds, it is a sign of a weak immune system. It means your pro-inflammatory immune pathway (known as Th1) is underactive, and your anti-inflammatory system (known as Th2) is dominant—a sign that your left brain may also be weak. If you almost never get colds but you have chronic allergies or sensitivities to all kinds of foods and chemicals, then the opposite is true: Your immune system is overactive. Your anti-inflammatory system is weak and your pro-inflammatory system is dominant, indicating a possible weak right hemisphere as well.

Typically when a doctor does an immune function blood test he or she is checking your levels of red and white blood cells. Elevated white blood cells are a sign of possible acute active infection somewhere in your body. Depressed white blood cells can be a sign of chronic infection. However, white blood cell count does not tell us much about the immune response itself.

A balanced immune system means there should be a balance between the two response systems. It is important to know if you have an imbalance because studies show that these two systems are off balance in children with autism and their parents. Your physician, functional neurologist, or functional medicine specialist can order a specialized test called the Th1/Th2 Cytokine Assay that measures the levels of immune chemicals called cytokines in each system. This will give an indication if levels are higher or lower than normal. It will also allow you to compare the two sides to find out if they are in balance or one side is more dominant. Though the test is not routine, it is inexpensive.

When the two systems are out of balance, it can lead to autoimmune issues or trigger allergies and food sensitivities you may already have. Eating certain foods or taking certain supplements can alter the balance of these two systems. So listen to your body if you enjoy such things as caffeine, red wine, or green tea, which are popular beverages that are known to trigger an immune imbalance.

I found this out myself when my allergies started acting up and I couldn't figure out why. When I looked at my diet I noticed that I had been drinking more red wine, green tea, and coffee, all of which increase the anti-inflammatory response. When I had my cytokine test done I found that my body was suppressing my pro-inflammatory response. This increased the sensitivity of my antibodies and I developed more allergies and was more sensitive to my environment. Allergies are related to inflammation so I suspected that my level of inflammatory chemicals was elevated. To correct this I eliminated coffee, green tea, and red wine from my diet and started taking supplements, such as echinacea. Almost immediately my allergies were gone. Not surprisingly, I discovered I also had a left-hemisphere weakness, which was also contributing or may have been the primary cause of my immune imbalance. It was also the reason why my allergies did not completely subside. Both improved when my left-brain function improved.

MAKING A CORRECTION

The only effective way to correct an immune imbalance is to first balance the brain, and then balance the rest of the body. You can begin this by completing the Melillo Adult Hemisphere Checklist, if you have not done so already, to find out if you have an imbalance and where it is. You can then help decrease an overactive immune response by doing the activities in Chapter 11 designed to help correct a brain imbalance. Because a heightened stress response can contribute to an immune imbalance, doing some of the stress release techniques recommended in Chapter 10 can also be helpful.

Certain foods and supplements can also help you correct an imbalance. If you cannot for some reason get the cytokine blood test, you can try experimenting with these foods and supplements. Make your best educated guess as to which system is too high and follow the protocol. If you don't notice a difference in a week or two or if your symptoms get worse, then stop and try the opposite. It is best to do this in conjunction with your doctor or functional neurologist. Do not take amounts of these supplements greater than recommended by your doctor or above what is recommended on the label. Also check to make sure that they do not interfere with any medications you may already be taking.

These Compounds Stimulate or Increase a Pro-Inflammatory Response and Dampen an Anti-Inflammatory Response

- Astragalus

- Beta-glucan mushroom

- Glycyrrhiza, which is found in licorice

- Echinacea

- Lemon balm (*Melissa officinalis*)

- Maitake mushroom

These Compounds Stimulate an Anti-Inflammatory Response and Suppress a Pro-Inflammatory Response

- Caffeine

- Green tea extract

- Grape seed extract

- Lycopene, the antioxidant abundant in tomatoes

- Pine bark

- Pycnogenol

- Resveratrol, the antioxidant in red wine

- White willow bark

You may notice that many of these substances are promoted by natural healers and nutritionists as immune boosters, and studies show they do work. However, in some people they may actually create an imbalance between the two systems. So, listen to your body. If you take any of these substances and notice signs of a weakened or overactive immune system, like I did, eliminate them.

Compounds That Modulate or Naturally Balance the Two Systems
- Probiotics

- Vitamin D

- Vitamin A

- Vitamin E

A substance in breast milk called colostrum also helps balance these two systems, which may be why breastfeeding has been found to be protective against autism.

CHAPTER TEN

Reducing Your Risk Factors

A Ten-Point Preconception Plan

All of my work is based on a functional neurology and functional medicine perspective, meaning that we must objectively measure before, during, and after whenever we are examining the brain or the body. The body is one coherent whole, so if ill health or an outside influence affects one system, it affects all systems, and that includes the brain. The body plays a role in brain function, so focusing on the brain alone when the problem *is* the brain is not the solution either. We need to measure and address all functions to be in optimal health and have the greatest opportunity to help lower the risk of autism.

We cannot assume anything. It is not adequate, and it certainly is not wise, to assume you or someone else has autistic tendencies, or a brain imbalance, or even that you are more left-brained than right-brained, or vice versa. It's not useful to assume that your problem is probably one thing but not another—that perhaps you are in pretty good health and all you need to do is watch your diet. There is only one way to know your true state of health, and that is to measure it.

This book builds up to what I see as ten major measurable facets of health that are relevant in your risk of having a child with autism:

1. Your overall health and, by extension, your brain balance, and the lifestyle choices you make to affect them.

2. Your ability to control the stress response.

3. Your immune health and the amount of inflammation you have in your brain and body.

4. The health of your gut and sensitivities to foods and/or food substances.

5. Your exposure to chemical pollutants and your body's ability to thwart heavy metals and other toxins.

6. The balance of sex hormones in your body.

7. How your body handles insulin and blood sugar.

8. Your activity level as measured by the amount of fat you are carrying around.

9. The real age of your body compared to your chronological age, as measured by your physical and mental fitness.

10. Your family history and genetic susceptibility.

All ten are important because, for the most part, they are in some way interrelated. You cannot put your attention on one in isolation of the others and expect major change. The only way to really know where you stand in relationship to all ten functions is by measuring. Addressing them all is relatively easy and can be approached with little expense. Many you can do on your own, but some require the assistance of a functional neurologist or a medical professional knowledgeable about brain science. As you go through this chapter you will find that there are "normal" reference points for all tests. Though these are important guide points, your most important comparison is yourself. The result of any medical test taken for the first time is called your baseline. For example, the results of your cognitive type test, the hemispheric checklist, and the brain balance assessment should be considered your baseline. They are what you will measure yourself against going forward. When you repeat the tests you will compare the results to your baseline to see if you are improving or regressing. A few of the tests recommended here are performed as part of a routine annual physical, so you should already have a baseline for them. If you don't know your baseline, ask your doctor. If for some reason this is impossible, you can always start keeping track of your baselines now.

If you find you do not fall within the normal range for any of the tests in this and the previous chapters, many of the practices and tips recommended in this book should help you see improvement. In some instances, such as finding a hormone imbalance or diagnosing an autoimmune deficiency, medical intervention is required. The important thing is knowing where you stand before taking remedial action. Keep track by recording the results of your tests in a special notebook. Again, an important point: No one abnormal measurement in and of itself means much in terms of your risk of having a child with autism. Risk is cumulative. The more risks you accumulate, the more alert you need to be to prevention. Here are measures you can take to track them.

1. Healthy Brain, Healthy Body

And I can say vice versa. Good health is a top-down operation, meaning it starts with the brain. I believe most of the issues addressed in this book are in some way associated with a brain imbalance. If you have too much of an imbalance between the activity of your brain's two hemispheres, communication and integration between the areas of your brain slow. How this manifests depends on the side that is impacted and the severity of the imbalance.

Bottom line: Autism appears to be related to a severe lack of integration in areas of communication between various networks in the brain and in particular between both sides of the brain, where the right side is too weak or immature or the left side is relatively more mature and too dominant. This abnormal growth starts early in life or in the womb. As I explained throughout this book, you may have autistic traits or imbalances you could pass on to your children that can increase their risk of getting autism. The way to measure this is by taking the quizzes in Chapters 5 and 6 that define your cognitive style and help you figure out if you have a brain imbalance. They are key, so if you haven't taken them yet, do so before moving on. These tests and the measurements you collect in this chapter are designed to give you a profile of your current state of health, both of body and brain. They all should be repeated periodically, and especially before you plan to have a child.

No matter where you are in your brain health, following the lifestyle recommendations in the next two chapters are important in your quest to improve your health and your brain balance. Read them and incorporate as many as possible into your life. Practice them as a family and establish them as a way of life. They are crucial to both healthy brain development *and* maintaining a healthy brain throughout life.

Also take the recommended measures to lower your stress response and keep your immune system healthy. I believe these are the biggest risk factors associated with autism, as they are in some way related to just about everything else that can increase your risk of having a child with autism. A high stress response and inflammation are systemic conditions now known to play a role in the development of many of the chronic diseases we are seeing today. This has become more and more apparent in recent years. Because their true role in disease has yet to be scientifically defined, they are not generally on the list of "must do" tests when your blood work is routinely checked during your annual physical. To this end, I strongly recommend both men and women have a comprehensive metabolic panel, which is a routine blood test, before starting to plan for a family. This test typically includes an analysis of numerous functions including the following:

- Blood sugar

- Cholesterol

- Electrolytes

- Liver and kidney function

- White blood cells

- Red blood cells

- Platelets

- Thyroid tests

A routine physical can but doesn't always check for other things that should be important to prospective parents. Ask your doctor to also include:

- A saliva test to check your cortisol level and your stress response

- Blood work to measure C-reactive protein, which measures inflammation

- Blood work to measure levels of vitamin D, which has become a common deficiency in the United States

TOP TEN LIFESTYLE PRACTICES FOR
PRECONCEPTION AND A HEALTHY PREGNANCY

In preparation for starting a family, you must be in the best physical and mental health. It will also prepare you for being an excellent role model to your children. To this end:

1. Make activity your number one priority. Unfortunately, too many people have sedentary jobs and hectic lives that push going to the gym or getting outside to exercise to the bottom of the to-do list. You *must* move it to the top. Get at least thirty minutes a day of aerobic exercise. You should also move around during the day instead of just sitting.

2. Eat for a healthy pregnancy. Both women and men should take prenatal vitamins at least six months before getting pregnant and make sure you eat foods containing these key nutrients. If you don't get them in your diet, consider the following supplements:

   - Vitamin D
   - Folate or folic acid
   - Vitamin B_6
   - Vitamin B_{12}
   - Vitamin E
   - Omega-3 fatty acids

3. Eat a clean diet. This means you should avoid sugar, eat only gluten-free grains, and put the emphasis on fresh organic green vegetables. Eat only high-quality organic lean meats in limited amounts. Get some protein at every meal and avoid sugar and all artificial sweeteners.

4. Soak up vitamin D. Sunscreen is a smart strategy to protect your skin from harmful ultraviolet light, but save it for the beach or when you are playing volleyball or other sports in the sun. More and more studies suggest getting vitamin D naturally is imperative

to a healthy pregnancy and in reducing the risk of having a child with autism. All it takes is fifteen minutes a day. Get out in the sunshine in the early morning or late afternoon when the sun is not so intense. Vitamin D production is the most efficient at times of the day when the sun is at a particular angle. Visit VitaminD Council.org for more information.

5. Practice stress reduction. If you have a stressful job, take up a stress-reduction technique and take time on the job to practice it.

6. Avoid stimulants such as coffee, tea, and all soft drinks. Limit alcohol consumption to red wine and, of course, no alcohol at all as soon as you start trying to conceive. Avoid recreational drugs.

7. Learn how to get a good night's sleep. Poor sleep is implicated in increasing the stress response and inflammation. If you sleep fitfully, meaning you do not feel rested in the morning, seek help to eliminate the problem.

8. Resist taking medication of any kind unless absolutely necessary.

9. Go organic. Avoid using pesticides in your home garden. Eat only fresh organic fruits and vegetables and wash them carefully.

10. Measure all the function for brain balance that you learned in this book regularly.

- A breakdown of cholesterol into high-density lipoprotein (HDL), the good kind; low-density lipoprotein (LDL), the bad kind; and very low-density lipoprotein (VLDL).

- And, if inflammation is found, or if you believe you have some autoimmune disorder, allergies, or immune dysregulation, a blood test to determine the imbalance of the pro-inflammatory and anti-inflammatory (TH-1 and TH-2) pathways.

You'll find out more about some of these tests and what to look for as you read on.

2. UNDERSTANDING AND MEASURING YOUR STRESS RESPONSE

When the concept of stress first became popularized in the 1960s and more widespread in the 1970s, the main focus was on the adrenal glands, which activate the release of stress hormones. At the time it was believed that the stress response went through stages based on acute and chronic stress that could eventually lead to what became known as adrenal fatigue. Many if not most nutritionists and physicians still believe that this is what happens. New research, however, shows that the problem isn't with the adrenal gland per se, but in the brain.

The stress response is the classic example of the brain's top-down regulation of body systems. The stress response runs on an axis that involves the pituitary gland, which is housed in the brain, and the hypothalamus and the cerebral cortex, which are both involved in the central nervous system. So, even though the adrenal glands sit far away atop the kidneys, the stress response is really mostly all about the brain.

The adrenal glands manufacture and secrete stress hormones on signals they receive from the brain. When certain areas of the brain are weak, however, communication between key areas of the brain and the adrenal gland misfires, making the adrenals kick out more cortisol. This is what raises the basic "idling speed" of the stress response.

The effects on mind and body are widespread. It wears on the immune system, making us more susceptible to infections and other problems, which lead to chronic inflammation. It can disrupt the imbalance of other hormones, including the thyroid and sex hormones.

The stress response is intimately related to your sleep-wake cycle known as your circadian rhythm. This is because the adrenal glands are also responsible for manufacturing cortisol, which regulates the cycle. Cortisol and melatonin are normally excreted in a specific pattern that mirrors your circadian rhythm. When one is high, the other is low. This rhythm allows you to feel awake and energized in the morning and during the day and to feel tired and ready for bed at night. If you have a chronically elevated stress response, it will eventually disrupt your circadian rhythm. If, for example, you get a midday slump and need a nap or you can't get out of bed in the morning, or have a hard time getting to sleep at night, these are all indications of a disruption in your circadian rhythm.

When you put it all together, you can see how a brain imbalance relates systemically to other problems in the body and how autism can involve them all. Kids with autism commonly share all these issues, including sleep problems.

Cortisol and Circadian Rhythm

Melatonin, which is the hormone that is secreted to help you fall asleep at night, has an antagonistic relationship with cortisol, which wakes you up. When cortisol is high, melatonin is low and vice versa. This is what we see with insomnia. If melatonin regulation is off, it will have a direct effect on cortisol regulation. It will also have a direct effect on your ability to get quality sleep. Lack of sleep in itself can cause an increase in the stress response, and an increase in the stress response can prevent you from falling asleep.

The amount of sleep we require is directly related to our body weight. Thinner people require less sleep and heavier people need more sleep. With that in mind, consider these common symptoms of disrupted circadian rhythm:

- Inability to fall asleep
- Inability to stay asleep

- Difficulty waking up in the morning

- Not feeling rested after a night's sleep

- Feeling tired rather than energized after exercise

- A drop in energy between four and seven p.m.

- Unexplainable surges in blood sugar

- A pattern of pain and headaches at certain times of the day

These factors can adversely affect melatonin levels and interfere with the sleep-wake cycle:

- Eating simple sugars and fats decreases oxygen supply to the brain, which decreases alertness and makes you sleepy.

- Drinking alcohol reduces the relative amount of time you spend in REM sleep. This is why you don't feel rested the next day after a night of partying, even when you get in a full eight hours.

- Food additives in general and artificial sweeteners in particular tend to increase alertness, which interferes with sleep.

- Eating a large meal in the evening, especially a high-carbohydrate meal, also interferes with sleep. It is best to have your last meal of the day no later than six p.m.

Symptoms of a problem with the adrenal glands' ability to properly secrete hormones include:

- Morning and/or evening fatigue

- Allergies

- Apathy

- Burnout

- Chemical sensitivity

- Depression

- Increased susceptibility to infection

- Insomnia

- Unstable blood sugar

- Low sex drive

The Brain Balance Way of Measuring Stress

Each side of the brain has neurological control of the stress response separately and independently. You can actually tell where this imbalance is by working your way down the body. These are all tests you can perform on your own that should give you a sense of whether you have a high stress response and the side of the brain that may be creating the problem. A medically supervised laboratory test can confirm if you have a high stress response. The following self-tests are designed to help you more accurately identify an elevated stress response in relationship to a brain imbalance.

Pupils
When the stress response is high, the pupils dilate and get bigger in order to see better. When the stress response is higher on one side of the brain and body, there will be a noticeable difference in the size of the two pupils. Normally, the pupil is larger on the side of the imma-ture or weaker brain.

Same in both eyes ____

Larger in right eye ____

Larger in left eye ____

Bilateral Blood Pressure

A blood pressure reading measures how much force is placed upon the artery walls as blood is pumped throughout the body. Blood pressure is measured in millimeters of mercury (mm Hg). The top number, called systolic pressure, measures the heart's contractions as it pumps blood from the heart. The bottom number, called diastolic pressure, measures your heart at rest between beats.

BLOOD PRESSURE REFERENCE RANGE FOR ADULTS

Normal:	120/80
Low:	90/60 and below
Prehypertension:	121/81 to 139/89
High blood pressure:	140/90 and above

More significant, however, is comparing blood pressure as measured in both arms. Common knowledge says that pressure in the left arm is higher because the heart is left of center. However, it is also known that more than a ten-point difference between the two arms is abnormal. Pressure equal in both arms or higher in the right arm are almost always considered abnormal.

Ruling out any major obstructions in a relatively healthy person, the most common reason for a significant difference in blood pressure in both arms is an imbalance in the brain and its control of the stress response. If the brain is weak and immature on one side, we would expect to see blood pressure higher in the arm of the same side. This is not common knowledge among traditional primary care physicians, who routinely take blood pressure only in one arm, usually the left.

I have been measuring blood pressure in both arms for years and I have taught thousands of professionals to do it the same way. It is a very accurate measure and tells you a great deal about your body and brain. Next time you are at your doctor's office, ask the nurse to measure the pressure in both arms.

Blood pressure in my left arm is _____

Blood pressure in my right arm is _____

The difference is _____ **points**

Pulse and Heart Rate

A high pulse and heart rate are frequently a sign of a brain imbalance. When the stress response is high, your heart rate and blood pressure will be high as well. Because these two measurements involve the heart and cardiovascular system, you won't see a difference from left to right as you do with blood pressure. Rather, they show up as a pulse and heart rate that are beating too fast.

As you will see from the following chart, heart rate slows as the brain matures. If the heart rate is high, it can be looked at as being more like an immature heart rate. It can also be a sign of a high stress response associated with a brain imbalance. This is especially true of men and women in their childbearing years.

REFERENCE RANGE FOR HEARTBEAT

Newborn to one month:	70 to 190 beats per minute
One month to one year:	80 to 120 beats per minute
Age one to ten:	70 to 130 beats per minute

Age ten and older, including seniors: 60 to 100 beats per minute

Well-trained adult athletes: 40 to 60 beats per minute

To measure your own pulse rate just put your finger on the pulse in your wrist or neck and, using a stopwatch or the minute hand on a clock, count the beats for six seconds. Multiply the number by ten. So if your heart beats six times in six seconds, your heart rate is 60. You're doing good!

My heart rate is _____ beats a minute

Breathing Rate

As the brain matures, the breathing rate comes down. Respiratory rate becomes more rapid and shallow when the stress response goes up. Respiration is measured at rest. Have someone assist you so you can lie still without distraction. All the person has to do is count the movement of your chest for one minute.

REFERENCE RANGE FOR BREATHING RATE

Birth to one year: 30 to 40 breaths per minute

Age one to three: 23 to 25 breaths per minute

Age three to six: 20 to 30 breaths per minute

Age six to twelve: 18 to 26 breaths per minute

Age twelve and older: 12 to 20 breaths per minute

My respiration rate is _____ breaths per minute

Sweat, Moisture, and Hair

We all experience sweaty palms now and then. Sometimes you can even feel sweat on your feet. They are classic signs of stress rising in your body. However, *always* having clammy hands and/or feet is not normal.

A chronically high stress response increases sweat production. When you feel moist and clammy all the time, the skin can start to look shiny and start to lose hair. It most likely will also be different on the two sides of your body. Check your underarm sweat stains on your clothing to see if one side is wetter than the other. Also notice the slipperiness in your hands from one to the other. Ask others to hold your hands and see if they notice a difference. Compare your hands to theirs. Notice if you have more sweat, moisture, or less hair on one arm and/or leg compared to the other. If you do, this is usually on the side of the brain that is weak and immature.

I do _____ do not _____ detect a difference in how I perspire

I perspire more on the L _____ R _____

Reaction to Cold

Hate cold weather? That's normal for many people. However, an exaggerated response to cold in which the blood vessels to your extremities constrict to the point where it is painful or they change color is not. A doctor most likely would diagnose this as Raynaud's disease. Raynaud's, like other autoimmune diseases, is often a sign of an overactive stress response. The condition can show up in both hands or both feet, but it can be more problematic when found on one side.

Raynaud's can occur on its own but it frequently shows up in conjunction with other autoimmune diseases, such as rheumatoid arthritis, lupus, scleroderma, or the thyroid conditions Graves' disease or Hashimoto's disease.

Some common triggers for Raynaud's include:

- Going outside during frigid temperatures
- Holding an iced drink

- Walking into an air-conditioned room

- Reaching into the freezer

- Putting hands under cold water

- Feeling emotional stress

I do ____ do not ____ have a sensitivity to cold or have been diagnosed with Raynaud's disease

Testing Stress

Because stress is so broad-based in its definition and yet so vague, measuring stress can be somewhat complex and perplexing. Part of the complexity is in understanding that there are three components to stress and how it is measured:

- Most people think of stress as a reaction to a traumatic experience or life event. For this, researchers have actually come up with a scale that rates life's events according to what has been found to have the greatest impact on body and mind. I gave an

---

DO YOUR FINGERTIPS WRINKLE?

Do your fingertips wrinkle after taking a bath or after having your hands in water for a long time? No problem, it's normal. However, if the fingers of one hand are significantly more prune-like than the other, it can be a sign of a high stress response. If you also notice other signs, such as excess sweating, on the same side of the body as well as other symptoms discussed in this chapter, then the likelihood is even greater that you have a high stress response. Get a cortisol test to find out for sure.

example of one such scale in Chapter 4. However, every person is different and physically reacts differently to stress, so in my mind this type of evaluation really doesn't tell you much.

- The second way to measure stress is to factor in an individual's personal differences by observing how they handle a situation and the environment they were in at the time. For example, this is something a psychologist is trained to observe and consider when ascertaining a person's mental state as a result of a particular incident or situation. This measure is done through professional observation as well as a patient questionnaire.

- The third approach is to measure your stress response by actually measuring the levels of the stress hormone cortisol circulating in your body. This is the most revealing and only true accurate measure.

We know through research studies that the actual event and even your emotional reaction can have little to do with how you respond biologically. A true professional diagnostic workup involves a slew of detailed and expensive tests involving virtually every system of the body. This type of workup is generally only recommended for someone with signs of a breakdown or post-traumatic stress syndrome. Though it is the best scenario to measure your true stress response, it is not practical or necessary for our purpose here.

If the self-tests I've just described as well as other tests you've already taken in this book indicate you may have a high stress response that may or may not be associated with a brain imbalance, getting your cortisol level checked is crucial. This is not something a doctor would typically consider in a routine visit, even if you are complaining about stress. As I said, general medical practitioners have a poor understanding of the connection between the brain and cortisol and how the brain can inhibit the sympathetic nervous system. However,

---

LIFESTYLE PRACTICES THAT REDUCE CORTISOL

Studies have found certain lifestyle practices can help bring cortisol levels down and reduce the stress response. They include:

- A healthy diet that includes soy foods, black tea, vitamin C, fish oil, and magnesium
- Mindful meditation
- Music therapy
- Sex
- Exercise
- Laughter

I believe the overall activity, maturity, and balance of the brain are what ultimately determine the level of anyone's physical response to stress, regardless of the level of the perceived stress. Someone who is under great perceived stress could have a very slight biological reaction with normal cortisol levels. On the other hand, someone who has very little perceived stress can have a significant stress response that may not even be in response to situation stress.

The Adrenal Salivary Index

There are three different lab tests that can assess the level of your stress response: a blood sample, a urine sample, or a saliva sample. Of the three, I believe the saliva test, known as the adrenal salivary index (ASI), will deliver the most accurate results because the blood and urine tests are too sensitive to pick up certain adrenal references.

The ASI is a home test. Your doctor or functional neurologist will give you a kit in which you will collect saliva four times a day over a twenty-four-hour period—morning, noon, midday, and at bedtime.

These are the times in which your circadian rhythm ebbs and flows. Your cortisol level should be highest in the morning, when you should feel most alert, and lowest at night, when you are ready to go to sleep.

Saliva is measured by what is known as nanomoles (nM). The reference ranges for the ASI are:

ADULT REFERENCE RANGE

Seven to eight a.m.:	13 to 23 nM
Eleven a.m. to noon:	4 to 8 nM
Four to five p.m.:	3 to 8 nM
Eleven p.m. to midnight:	1 to 3 nM

3. Tame the Flame

Inflammation is not good. In an ideal world, inflammation exists to clean out infection and then subsides in order for healthy tissue to rebuild. This is what is known as acute inflammation. If an infection is stronger than the immune system, inflammation keeps burning and becomes chronic, hanging around for weeks and possibly even years.

Inflammation can infest the body and brain for reasons that have nothing to do with infection. Chronic inflammation is associated with a number of health issues, most of which at one time were considered to be "diseases of aging." But not anymore. We are seeing age-related diseases such as heart disease and diabetes at a younger and younger age. Sadly, we are even seeing signs of these conditions in children.

Studies show that overweight and obesity—specifically too much fat around the middle—is one of the leading causes of chronic inflammation. It also raises a woman's risk of having a child who will get autism. Research suggests that the body reacts to overweight with an

LIFESTYLE PRACTICES THAT
INCREASE CORTISOL: THE BIG THREE

On the surface, a high cortisol level doesn't appear to fit as "a lifestyle practice that increases the risk of autism." However, lifestyle practices are what elevate cortisol levels and keep them chronically high. The big three:

- A sedentary lifestyle with little or no exercise
- A diet containing pro-inflammatory (starch and sugar) foods along with inadequate levels of essential vitamins, minerals, and anti-inflammatory antioxidants
- A high percentage of body fat, especially around the middle

Studies have found these conditions and practices can trigger a high stress response:

- Burnout
- Drinking caffeinated beverages
- Daily commuting, with levels rising on the length and difficulty of the trip
- Intense and prolonged physical exercise
- Inadequate sleep and poor sleep quality
- Severe trauma or a stressful event

Reverse the Response with Supplements

There are herbal supplements called adaptogens that can help reduce or normalize the stress response as needed. Also, supplements containing a substance called phosphatidylserine has also been found to help lower the stress response. Getting high levels of anti-oxidant vitamins and glutathione through food and supplements is very important to reduce the effects of a high stress response. You can find more information about these supplements and how to purchase them in the Resources section on page 302.

autoimmune response, mistaking fat deposits for intruders and attacking them as if they were a bacteria and fungi. I believe, and the research suggests, that the real cause is the chronic inflammation that travels with obese people. Scientists believe that excess inflammation in the mother's blood is transferred through the placenta to the fetus and affects the developing brain. If you are carrying around too much fat, you most likely have chronic inflammation.

Inflammation generally can also be detected if you have allergies, arthritis, diabetes, metabolic syndrome, thyroid disease, or any of the myriad autoimmune diseases. Its role in cardiovascular disease is still unclear, but some experts believe it may be an even more important marker than high cholesterol and high blood pressure. This is based on documentation showing people have had heart attacks in the absence of all other known risk factors *except* inflammation.

Symptoms and chronic conditions that are associated with chronic inflammation include:

- Allergies

- Arthritis or painful joints

- Chronic fatigue syndrome

- Chronic headaches

- Diabetes

- Digestive problems

- Fibromyalgia (muscle pain and fatigue)

- Heart disease

- Metabolic syndrome

- Eczema

- Environmental and chemical sensitivities

- High blood pressure

- High cholesterol

- Irritable bowel syndrome

- Overweight and obesity, specifically high body fat around the waist

- Prostatitis

- Rosacea

- Sleep apnea

- Short-term memory problems

If you have any of these conditions, you most likely have chronic inflammation. However, if you don't have any of these problems, you still cannot rule it out. The only way to know is to test.

As we get older, the mechanisms that control inflammation become more sensitive, making the immune system more vulnerable to the environment and outside invaders. In fact, more and more evidence suggests that chronic inflammation may be a reason why certain health problems develop. I believe chronic inflammation is the reason why we are seeing more autoimmune conditions today and in younger people.

Testing for Inflammation

Low-grade chronic inflammation is characterized by a 200 to 300 percent increase in concentrations of circulating cytokines, substances in the blood that are markers of inflammation. Because chronic inflammation as a risk factor in cardiovascular disease is the subject of

LIFESTYLE STRATEGIES TO CONQUER
STRESS AND INFLAMMATION

A preponderance of research shows there are three major ways to tackle stress and chronic inflammation and all the ramifications of health that are associated with them:

- **Eat a proper diet.** This means a natural, preferably organic, diet devoid of food additives, high salt, and simple carbohydrates. Most of all, it means a weight-loss diet if you are overweight.
- **Exercise, exercise, exercise.** There is no better way to relieve stress than through good heart-pumping exercise. Tai chi and yoga are excellent disciplines well known to bring down the stress response.
- **Learn to be mindful.** There are dozens of stress reduction techniques that can help bring down the stress response, including just learning to relax. However, the most well-documented and effective way to bring down the stress response is through a practice called mindfulness, a type of meditation that involves taking a few minutes of time throughout the day to erase all thought from your head and just breathe. Mindfulness meditation can be learned on your own, but it takes practice to master it. There are many good books on the subject, including those by Jon Kabat-Zinn, MD, the founding director of the Stress Reduction Clinic and Center for Mindfulness in Medicine, Health Care, and Society. Mark Bertin, MD, a developmental pediatrician, tells how to approach mindfulness from the perspective of a parent with a child with a developmental disability in his book *The Family ADHD Solution*.

intense research, many doctors now routinely check for it in blood work performed during your annual exam. If your doctor does not routinely check it, request that it be included. The test, which is called C-reactive protein or a CRP blood test, is simple and inexpensive. Its reference range is measured according to your risk for cardiovascular disease.

Testing for inflammation is one of the most important must-dos for anyone considering pregnancy. Both men and women should take all necessary precautions to avoid or minimize inflammation. The outcome you are looking for in the test is to have no detectable CRP. A detectable level of CRP means you have inflammation.

C-REACTIVE PROTEIN REFERENCE RANGE

Less than 1 mg/L:	low risk
1 to 3 mg/L:	medium risk
Greater than 3 mg/L:	high risk

If the test shows you have inflammation and you also have an autoimmune condition or are plagued by allergies and stomach problems, you should also investigate the balance of your immune system, which I detailed in Chapter 9. This too is a simple blood test called the Th1/Th2 Cytokine Assay.

4. Focus on the Stomach

When inflammation looms, one of its favorite places to burn is in the gut. Chronic inflammation and leaky gut syndrome go hand in hand. If you test positive for chronic inflammation, you likely also have chronic stomach trouble and food sensitivities.

Leaky gut means you have a weak stomach lining, which offers an escape route to molecules that would generally be too large to leave the stomach. In essence, it opens the door for unfriendly residents, such as undigested proteins from food, yeast, fungus, bacteria, parasites, and toxins, to roam the bloodstream and set up a hostile environment. Once in the bloodstream, the immune systems goes into attack mode, calling on the body's detoxification system for help. Eventually this weakens the body's ability to properly detoxify, causing sensitivites to certain foods and substances. This, in turn, can cause more inflammation and create a vicious cycle.

If you have indications of chronic inflammation and a leaky gut, I suggest you seek out a professional trained in functional medicine or functional neurology who can evaluate you for these infections and help you eliminate them. Once the problem is identified, your doctor most likely will employ a professionally supervised protocol called the Four Rs Program—remove, replace, reinoculate, and repair—to restore health to the gut and stop inflammation.

Testing for Leaky Gut Syndrome

There are presently two tests for leaky gut. The most common is a simple urine test called the lactulose and mannitol test. After an overnight fast you will be asked to drink a mixture of these two inert sugars and your urine will be tested over the next six hours. Mannitol, a small molecule, is easily absorbed but lactulose is too large to absorb—at least in a healthy stomach. As urine is collected, the ratio of the sugars being absorbed is measured.

REFERENCE RANGE

Manitol:	14 percent
Lactulose:	1 percent

Recently a newer and more accurate way of testing for leaky gut has been developed by Aristo Vojdani, PhD, the chief scientific advisor for Cyrex Laboratories and one of the world's leading scientists in immunology. Cyrex specializes in development tests for autoimmune

STOP INFLAMMATION WITH THE 4 RS

Remove, replace, reinoculate, repair. These are the steps to the 4 Rs Program, a tried-and-true protocol that restores health to the gut and reduces inflammation. It is one of the foundational principles of functional medicine and I have found it works really well. If you have chronic inflammation, I recommend you find a physician or functional neurologist versed in the 4 Rs Program. It is not something you can do or should try on your own.

It takes about twelve weeks to work through the program and it is modified to suit your needs. These are the four steps:

1. **Remove.** In this phase you remove all substances from your diet and environment that might be disrupting the immune system and creating inflammation. Focus is on food sensitivities, so all common allergy-producing foods, including grains containing gluten and dairy products, are eliminated from the diet. A diet with low-allergy potential is prescribed, which encompasses rice-based products, legumes, fruits, vegetables, fish, and poultry. Exposure to toxic chemicals and toxic metals, such as mercury, are also eliminated.

2. **Replace.** This phase focuses on digestive enzymes and proper stomach acid. It looks to see if you can properly digest a normal meal without complaints of bloating, gas, or reflux. If this is a problem, digestive aids, such as enzymes, will be prescribed.

3. **Reinoculate.** In this phase you are prescribed a therapeutic dose of a combination of probiotics, usually consisting of three to five billion live culture organisms, designed to improve intestinal immune function, and thereby improve overall immune function. Probiotics are specific strains of acidophilus and bifidus bacteria that are normal inhabitants of a healthy intestinal tract. Most likely, you will also be asked to take probiotic supplements.

4. **Repair.** This phase consists of nutritional supplements to promote proper repair of the intestinal lining. Supplements generally include the amino acids L-glutamine and glycine, pantothenic acid (vitamin B$_5$), vitamin E, fish oil, and zinc citrate.

---

disorders. Dr. Vojdani's test is a specific blood test indentified as Panel 2 that you can order and take yourself through cyrexlabs.com or you can request from your doctor or functional specialist.

Looking for Food Intolerance

As I discussed in Chapter 4, having a food intolerance, or what doctors call a nonallergic food sensitivity, is a lot different than a food allergy. They bear no relationship to each other. A food sensitivity is more chronic, less obvious, and much more difficult to identify than a food allergy that shows up immediately, often with life-threatening consequences.

A food sensitivity is much more insidious. People can live for decades without ever knowing they have one. All they know is that they "aren't feeling right," though they generally do not connect it directly to food because it can happen hours and sometimes even days after the offending food is eaten.

I believe that the vast majority of food intolerances are caused by three things:

- An imbalance in the immune system, making it overly sensitive

- Dysfunction of the digestive system and the gut

- Poor regulation of the digestive system and the immune system by the brain

Symptoms can vary greatly from person to person, confounding the problem even more. They can include but are not limited to:

- Gastrointestinal upsets, such as gas, indigestion, constipation, and/or diarrhea

- Irritable bowel syndrome

- Inflammatory bowel disease

- Canker sores (mouth ulcers)

- Respiratory problems, such as chronic nasal congestion or a dry cough

- Skin eruptions, including inflammatory conditions such as dermatitis and eczema

The standard way to find out if you have a food sensitivity is through a blood test that measures a specific antibody that is released when you eat the offending food. The test is called an IgG food intolerance test. IgG stands for immunoglobulin G, one of the five subclasses of antibodies the immune system releases to chase specific antigens. (IgE tests for food allergy.) If levels of IgG are high, it means you are sensitive to a substance your immune system sees as an enemy. The reference range for IgG may vary based on the specific test and the laboratory used. Another way actually tests white blood cell response; this is known as an ALCAT test and is another good way to measure food sensitivities.

Knowing you have a food sensitivity is one thing, but finding out what it is you are sensitive to is something else. It sometimes can take a lot of detective work.

If you and your doctor have a hunch what the offender is, he or she may have you take a hydrogen breath test. This is a noninvasive test in which you eat a small amount of the specific substance after an eight-

or twelve-hour fast. The technician then measures hydrogen levels in the breath every fifteen, thirty, and sixty minutes over the course of several hours. Food intolerance often, but not always, produces hydrogen gas, so a negative test means that you might have to keep on searching.

Testing for Gluten

Though testing for food sensitivites in general can be done through a blood test, there is only one test that I can recommend to detect gluten sensitivity. This test also comes from Cyrex Laboratories, and it is called Panel 4. It is the most complete and accurate test because it examines all aspects of the gluten molecule, rather than just one, as most other lab tests do. Any part of the gluten molecule can trigger an autoimmune reaction. Only by testing all of them can you know for sure if you have a gluten sensitivity.

It has clearly been shown that gluten sensitivity is much more common today than it has been in the past. One excellent study compared blood tests taken from soldiers in the 1950s at Warren Air Force Base in Cheyenne, Wyoming, and to present-day soldiers. The researchers concluded that celiac disease, the most severe form of gluten sensitivity, has been on the increase over the past several decades. Studies also show that sensitivity in mothers is associated with an increased risk of having a child who will develop autism.

Array 3 also tests for cross-reactivity with other substances, such as dairy, soy, and other common sensitivites. I recommend getting this test along with the other food sensitivity tests to get the complete picture of your dietary problems.

The Elimination Diet

The test most likely to produce real results is an elimination diet. This is a rigorous diet in which all suspected foods are removed from the diet for four weeks, then gradually reintroduced one by one until a set of symptoms identifies the offending food. The food elimination diet has been around for decades. Nutritionists use it all the time in an effort to get to the bottom of a puzzling dietary dilemma.

You can do this diet on your own, but for best results I recommend doing it under the supervision of a knowledgeable nutritionist, functional medicine specialist, or functional neurologist. It's a long process and can get tedious because you'll most likely crave the foods you are trying to avoid. The fact that you are working with a professional, and therefore you're accountable to someone, should help you stick with it. The diet consists of these steps:

Step 1: Round up the usual suspects. Of course you don't know what you are sensitive to, but you surely have an idea. Often we are sensitive to the foods we are "addicted" to. It's part of the vicious cycle of inflammation and a sensitive gut. We crave what our body rejects. The most common food sensitivities are to wheat (gluten) and dairy products (casein). These two foods also are chemically very similar, so it is very likely that if you discover you have a problem with one, you also have a problem with the other. Other common suspects include:

- Apples

- Baker's yeast

- Brewer's yeast

- Chocolate

- Corn

- All dairy and milk products (casein), including goat milk

- Eggs

- Legumes (beans, peas, peanuts, soy)

- Oranges and all citrus fruits and juices

- Refined sugar

- Soy

- Tomatoes

- Anything containing wheat or gluten

Step 2: Out with the bad, in with the good. You go cold turkey for the next four weeks and completely eliminate all suspected foods from your diet. If you had a blood test or tests that identified food sensitivities, eliminate all of these foods. However, you must also eliminate all unhealthy foods from your diet. This includes:

- Junk food—this means all fast food, candy, soft drinks, etc.

- Processed foods—this includes pressed meats and cheeses and most fast-to-the-table foods that come packaged.

- Food additives—this requires checking labels carefully. The list of food additives is long and you should get familiar with what they are. They include any ingredients containing the words *agents, enhancer, regulator, gums,* and ingredients ending with "ant."

Be aware that you may go through withdrawal symptoms as a result of eliminating some of these foods. However, in the end you will feel a measurable difference in your health, mood, and mental outlook. Be aware that giving in to a craving means you'll have to start all over again. So bite the bullet and stick with it, even if you notice these common "withdrawal" symptoms:

- Irritability

- Depression

- Lethargy

- Difficulty sleeping

Step 3: Welcome back the offending foods. In this step you begin reintroducing the foods on your suspect list one by one. What you start with does not matter. However, you should not do this step unless you have followed a healthy diet for four weeks. Reintroduce them in this manner:

- Eat the food at breakfast, lunch, and dinner, increasing the amount with each meal.

- Record the quantity of the food eaten and the time.

- Record all the symptoms you observe, the time they occurred, and for how long.

- At the beginning of each day, make a note as to the quality of your sleep the night before and your general behavior the next day.

- Reintroduce the food for only one day and then remove it again from the diet, even if it does not produce any symptoms. Symptoms can vary; however, whatever symptoms you had or felt before eliminating the food may return with reintroduction.

- Wait three days because it can take that long for symptoms to show, then begin reintroducing other foods on the list, one by one. Wait three days between each new introduction until all the foods have been tested.

- If you get a cold or infection during this step, suspend the process until after recovery.

- Milk should be considered and tested separately from other dairy products.

- If a food challenge results in significant symptoms, do not continue the challenge until all the symptoms are gone for twenty-four hours.

5. Searching for Signs of Environmental Chemicals

The brains of infants and children are especially vulnerable to environmental exposures. It is no coincidence that most children who have autism are sensitive and susceptible to environmental chemicals from the moment they are born. I believe this is largely due to starting life with a brain imbalance and exposure to toxins in the womb.

As you already learned, new research in neuroscience has identified critical windows of vulnerability in fetal life and in early childhood when exposure to certain toxic chemicals and medications can have a devastating impact on the developing brain. Toxins can harm fetal and young brains at levels much lower than what are considered safe in adults.

The role environmental chemicals play in increasing your chances of having a child with autism, however, is more insidious than just breathing in toxic chemicals. My concern is not so much toxic exposure itself, but the mother's ability to detoxify what she is being exposed to. Studies suggest fetuses are most vulnerable in mothers who cannot adequately detoxify harmful chemicals. Again, we see a vicious cycle. People with impaired detoxification systems almost always have digestive problems, inflammation, poor immunity, high blood sugar, and a hormone imbalance. On the other hand, a body and brain in good health have a good detoxification system.

Correcting the problem is not as simple as putting yourself in a detoxification program. You have to correct the source of the problem, which in many cases may be a brain imbalance. This is why you shouldn't try to tackle this problem by yourself and you should enlist the guidance of a functional neurologist.

The list of potentially harmful toxins is increasing as time goes on, and I am sure it will have expanded beyond those already identified by the time this book is in print. We can't escape environmental chemicals, but we can do our best to avoid them. The only way to know your personal exposure is by testing. I believe testing for levels of harmful or potentially harmful chemicals, heavy metals, and medications that could be harmful to the developing brain is something every man and woman should do before conceiving. Women should continue testing periodically throughout pregnancy. Testing itself is not harmful and only involves a simple blood test to detect antigens that chase harmful substances. If high levels are detected, a qualified professional can recommend a safe and effective detoxification program.

Your body detoxifies toxins naturally by releasing a substance called glutathione. Levels of glutathione, combined with the balance of your brain and nervous system, help determine your ability to purge toxins from your body. Unfortunately, there is no good way to test levels of glutathione in the body. However, if you have a brain imbalance, inflammation, leaky gut, or a compromised immune system, then most likely your glutathione level is low.

There are two ways to increase your levels of glutathione. One is with a topical supplement called liposomal cream that can be rubbed on the skin. However, I believe an even better way is by taking supplements called glutathione recyclers. Because there is no way to measure

AVOIDING PESTICIDES

Here are some tips to help you avoid pesticides and other chemicals in your food without sacrificing nutrition:

Go organic. This is first and foremost and applies to your home garden and your supermarket. Yes, it is more expensive, but it is well worth the cost when you consider what we know about harmful pesticides and what we don't know yet, which is more than we do know.

Wash, wash, wash. Thoroughly wash *all* fruits and vegetables, even if they are organic. Tap water is fine but a commercial produce-washing product is even better. Keep in mind, however, that even though chemicals can be washed off the surface of fruits and vegetables, they do seep into the soil that nurtures plant food, meaning they can get into our food systemically. This is another reason why it is better to buy organic.

Avoid the dirty dozen. If you can't buy organic, avoid the conventionally grown "dirty dozen" that are known to contain high levels of pesticides:

- Apples
- Bell peppers
- Celery
- Cherries
- Grapes
- Nectarines
- Peaches
- Pears
- Potatoes
- Raspberries
- Spinach
- Strawberries

glutathione accurately, I suggest taking these supplements preventatively. There are many options for removing harmful or potentially harmful chemicals from your body. A comprehensive detoxification program is, in my mind, the first line of action. One such protocol is the 4 Rs Program that is explained in the box on page 242.

Chelation therapy is another option. Chelation is a form of therapy in which substances are taken orally or intravenously that bind with heavy metals and other toxins and remove them from the body through the urine. Chelation, however, should be considered your last phase of rebuilding your health. Chelation should only be used in cases of acute toxicity, such as lead or mercury poisoning. These conditions, characterized by high levels of the substance in the blood, are rare. Chelation should not be carried out until the brain is balanced, the gut is healed, and the blood-brain barrier is working properly. These protective barriers need to be intact before attempting chelation, or the problem might get worse. If the brain is not working to control the digestive system, the immune system, hormones, and blood sugar, the entire detoxification system could break down. All of these systems are dependent on one another. Metametrix is a lab company that has joined with Genova Labs. They have developed a test for environmental toxins that seems to be among the best available at this time.

6. Balancing Hormones

One of the many ramifications of a high stress response is the potential to alter levels of testosterone in women and estrogen in men. Some studies suggest and several researchers believe that elevated levels of the male hormone testosterone contribute to autism.

Why testosterone is high in certain women is not clear, but I believe it is involved in a high stress response, high perceived stress, a brain imbalance, and poor lifestyle choices. For example, studies have found that amniotic fluid collected from pregnant women undergoing amniocentesis show high testosterone levels correlate with high cortisol levels. Also, studies have tied high testosterone levels to increased cortisol and surges in insulin. In women, the combination of increased cortisol and insulin surges releases an enzyme that converts estrogen to testosterone. In men, the opposite happens. A different enzyme is released that converts testosterone to estrogen. Increased estrogen in men and increased testosterone in women are seen as a result of a chronically high stress response and can also increase the stress response and create the vicious cycle that ultimately may raise the risk of having a child with a disability such as autism.

For these reasons, I believe testing hormone levels, especially testosterone and estrogen, is very important for men and women who are planning to have a child. The test is easy, inexpensive, and can even be done at home. I recommend women have the test performed prior to conception and throughout pregnancy.

The three basic ways of testing are through blood, urine, and saliva. Blood tests are commonly used in a doctor's office; however, they are somewhat unreliable because they cannot detect the 5 percent bioavailable hormones that are utilized by the brain, uterus, breasts, and prostate. The saliva test can. The World Health Organization started recommending saliva as the test of choice more than five years ago. Tell your doctor that this is the test you want.

Because hormone levels fluctuate throughout the day, the test is performed several times over a twenty-four-hour cycle. There are many kinds of hormone-testing kits, so make sure you use one that tests for these hormones:

- Estrogen

- Progesterone

- Testosterone

Timing of the test matters. Women should take the test during the latter half of the menstrual cycle. This would be any time during days nineteen and twenty-one, with day one considered the first day of your period. Men who are not taking hormones can have the test done at any time. If you are a man taking hormones, consult your doctor about when the best time is to have the test.

The reference range for hormone testing is complex, so it is best to discuss this with your doctor. Depending on the results, a functional medicine doctor or functional neurology should be able to advise you on a diet and supplement program that can help get your hormone levels in balance. There are several labs that do this type of hormone testing; Metametrix, Genova Labs, and ALCAT Worldwide are just a few.

7. Balancing Blood Sugar

Twenty years ago the term *prediabetes*—meaning you don't have diabetes yet, but you are getting mighty close—didn't exist. Today seventy-nine million Americans are walking around with elevated sugar levels, and many of them are men and women in their child-bearing years.

High sugar, or what we call insulin resistance, in a mother-to-be is not good. It has been implicated as a major risk factor in autism and contributes to problems during pregnancy that can result in risky delivery. It also affects virtually every other system in the body.

When the pancreas has to work overtime in response to a poor diet, cells begin to get insensitive or unresponsive to insulin, meaning the pancreas has to churn out even more hormones to get the job done. This disrupts adrenal function, which releases cortisol and turns up the stress response. It causes estrogen to convert to testosterone in women or the reverse in men, tipping the balance of these two important hormones. It irritates the immune system, which causes inflammation. It weakens and inflames the digestive tract and disrupts the liver from doing its job of detoxifying harmful toxins. The organ it affects the most, however, is the brain. The brain depends on insulin more than any other organ because it depends on blood sugar for energy. It's all part of the same vicious cycle of ill health breeding more ill health.

Making sure your blood sugar is normalized is critically important in preparing for a healthy pregnancy. Dysglycemia is an umbrella term used to define any abnormality in blood sugar. You should be aware of the symptoms. Hypoglycemia is a serious condition in which blood sugar is too low, which can cause you to pass out and go into insulin shock. It can be a life-threatening condition. Symptoms include:

- Cravings for sweets

- Irritability and/or light-headedness when meals are missed

- Blurred vision

- Poor memory

- Out-of-character emotional upset

- Caffeine dependency for energy

- Eating to relive fatigue

- Jitters

Elevated blood sugar is called hyperglycemia. Symptoms include:

- Increased thirst

- Frequent urination

- Fatigue after meals

- Exhaustion

- Constant hunger

- Cravings for sweets that are not satisfied even after eating sweets

- Strong desire for sweets after a meal

- Traveling aches and pains

- Large hip-to-waist ratio

- Difficulty losing weight

Measuring Sugar

Blood sugar can be measured through a blood test, but the easiest way to continually monitor your blood sugar level is to purchase a glucometer from your local pharmacy. It is relatively inexpensive and easy to use at home. You should check your blood sugar first thing in the morning before eating or drinking anything other than water, and then test it again two to four hours after each meal. Blood sugar is measure in milligrams per deciliter (mg/dL).

BLOOD SUGAR REFERENCE RANGE

Normal:	80 to 100 mg/dL
Low blood sugar (hypoglycemia):	Under 80 mg/dL
Insulin resistance (prediabetes):	100 to 125 mg/dL
Diabetes:	126 mg/dL and above

There are dietary measures that can help you get blood sugar under control and maintain a healthy level. Your doctor or a nutri-

LIFESTYLE STRATEGIES TO
MAINTAIN PROPER BLOOD SUGAR

- Eat a high protein breakfast
- Eat a small amount of protein every two to three hours
- Get a glycemic index chart and eat primarily low glycemic foods
- Do not eat high-sugar foods without fiber, fat, or protein
- Do not eat sweets or high-sugar foods before bedtime
- Avoid all fruit juices and carrot juice
- Avoid stimulants
- Eat a well-balanced diet consisting mostly of lean meat and vegetables
- Eliminate foods that you are sensitive or allergic to

tionist can help you. An important aspect is to follow a low glycemic diet. All foods are rated on a scale called the glycemic index, which is based on how rapidly a food causes a rise in blood sugar. The index is available from your doctor and on numerous sites on the Internet.

8. WHAT'S YOUR BODY SHAPE?

There is one common denominator in virtually everything discussed in this chapter: body weight. Entering pregnancy at a healthy weight is crucial to a healthy pregnancy and giving birth to a health baby. There are just too many risks involved in being overweight or obese to see it any other way. Of all the things a parent can do to help control all the risk factors involved in autism, I believe maintaining a healthy weight is at the top of the list.

Body fat is essential to healthy living, but only to a degree. It is necessary to maintain life and to keep reproductive organs functioning properly. Because of hormones and the demands of childbearing, the percentage of essential body fat for women is greater than that for men:

BODY FAT PERCENTAGE REFERENCE RANGE

Description	Women	Men
Fat essential for living	10–13%	2–5%
Athlete	14–20%	6–13%
Fit individual	21–24%	14–17%
Average person	25–31%	18–24%
Obese person	32%+	25%+

Anything higher than what is considered normal for you is stored fat, the kind many of us try hard to avoid. When considering body fat, the concern is not what we can see, such as a jelly belly or a big butt. It's the fat we can't see—the fat that surrounds our internal organs. This is called visceral fat, or visceral adipose tissue. Studies show visceral fat, or rather the lack of it, defines our health status. This means it is not your weight per se that matters the most, but the amount of fat you are carrying around, especially around the middle. This is why measuring your body fat percentage rather than just measuring your weight is the best way to judge your actual fitness. Someone could be two hundred pounds and have 30 percent body fat or be two hundred pounds and have 10 percent body fat. The first person would look obese and grossly out of shape; the other would be a perfect physical specimen, even though they may both be the same height and weight. Research shows that people who are apple shaped, with more fat around their waist, have more health problems and more health risks than people who are pear shaped and carry more fat around their bottom. It is quite possible to be average weight or even underweight and still have too much body fat. While knowing your body mass index (BMI) can give you a good idea where you stand, a better indicator is your hip-to-waist ratio.

To find your hip-to-waist ratio, stand relaxed and, using a stretch-resistant tape measure, first measure your waist just above the navel and then your hips at their widest part. Do the measurements twice to ensure accuracy. Divide your waist measurement by your hip measurement. For example: 28-inch waist divided by 38-inch hips equals 0.74.

REFERENCE RANGE FOR HIP-TO-WAIST RATIO

Men:	0.90
Women:	0.80

Ideally you want to have a ratio below the drawing line. A ratio of 0.80 for women and 0.90 for men correlates to generally good health and fertility. For women, it also indicates optimum levels of estrogen, and a reduced risk for diabetes and cardiovascular disease.

Body Mass Index

You should also know your body mass index. BMI is the measurement science generally relies on to measure weight and its relationship to disease. It is based on a complicated formula as follows: weight divided by height squared, multiplied by 703.

BMI REFERENCE RANGE FOR MEN AND WOMEN

Normal:	18.5 to 24.9
Overweight:	25 to 29.9
Obese:	30 and above

Your goal, especially if you are a woman, is to aim for the lower end of normal. A 24, for example, puts you uncomfortably close to many of the healthy risk factors discussed in this book. By the same token you don't want to be below 18.5. Being underweight carries its own set of health risks.

Admittedly, BMI has its shortcomings. Due to differences in body composition, BMI is not necessarily an accurate indicator of body fat percentage. It does not factor in bone structure or whether you are male or female. Also, individuals with greater than average muscle mass will have a higher BMI. The thresholds defining the line between "normal," "overweight," and "obese" are sometimes disputed for this reason. There are a variety of other ways to get an accurate measurement of your body fat, but these tests can be costly and generally are not covered by insurance without a good medical reason for doing so. However, keeping track of your BMI is relatively easy. All you need to know are your weight and height and all you need to have is the chart. BMI charts are readily available in a number of places online as well as in your doctor's office.

Many home weight scales now include a measure of body fat percentage. Also almost any chain pharmacy will carry a handheld body fat analyzer; you can even order one online. It is inexpensive and very easy to use. I tell almost every person I work with to get one. This method, of course, is not as accurate as some of the others listed, but it is accurate enough.

9. The "Real" Age Factor

You would train for a marathon, you would study for an exam, so you need to train to get pregnant. It's a team effort! If you go through all the tests I laid out in this book, you should have a pretty good indication of your health status. But how does it compare to where you should be at this point in your life?

As you know, advancing age of the mother and father plays a major role in the risk of having a child with autism. I believe, however, that age is just a number. It's quite possible the true age of your body has nothing to do with your chronological age. There is a way to find out. It's called the Real Age Test (realage.com) and it was developed by America's best-known physician and surgeon, Dr. Mehmet Oz, and his partner, Dr. Michael Roizen. More than twenty-seven million people have taken part in the test since it was first put online in 1999.

Dr. Oz says the calendar only reflects your chronological age, but the Real Age Test reveals your true physiological age. So even though the calendar says you're forty, you could actually be much younger—or older. A recent study that measured testosterone levels in men in their forties confirmed this. Conventional wisdom says that testosterone decreases with age in all men. However, the researcher found that this was not the case in men who were in very good physical condition.

Dr. Oz's test is very revealing and I believe it is very important to anyone who wants to have a child. It puts you through a series of questions about your lifestyle practices, your exercise and eating habits, and your general health. No matter what your age, you should take it. If you are only thirty but the test reveals your body is really closer to forty, then you know you have some work to do before you consider pregnancy.

Another way to think of your real age is to use yourself as a comparison. Think back a decade or so. What were you capable of doing when you were twenty, twenty-five, or thirty? How much do you weigh now and what is your percent of body fat compared to ten years ago? How about your blood pressure, heart rate, and blood sugar count? How fast can your run, how far can you go, how many push-ups can you do compared to your younger days?

FATTY ACIDS: A BRAIN BALANCE LINK

Not all fat is bad. In fact, fat is essential to brain health. Fat is necessary for cell membranes, nerve coverings, hormone production, vitamin absorption, and more.

Most people, including kids, get too much fat in the diet—mostly the wrong kind. Evidence is mounting that there is a link between low levels of fatty acids and the incidence of both ADHD and autism.

Good fats in the form of omega-3, -6, and -9 fatty acids are found in foods such as cold-water fish, flaxseed, and olive, vegetable, and nut oils.

Your goal is to get your body and mind to an ideal age range and keep it there as long as possible.

10. Know Your Family History

We know from the results of family and twin studies that genes play a crucial role in the development of autism as well as other neurodevelopmental disorders. If you have certain genes or certain copies of genes, the environmental factors discussed in this book can play a role in how these genes will or will not be expressed and passed on to your children and possibly future generations. You've read how this can happen.

Having one or even all of these genes does not mean you will have a child with autism. It is just another factor that raises the risk. In the end it is how these genes interact with other brain-building genes, your environmental exposures, and the lifestyle choices you make that will ultimately determine how these genes will be expressed, or if they will be expressed. For example, we know that the MTHFR gene that interferes with absorbing folate is a risk factor for autism, but we also know that taking prenatal vitamins in advance of pregnancy and throughout pregnancy can ward off the deficiency. For every factor and every gene, there is likely some avoidance factor. In autism, it appears that nothing is inevitable.

That said, you may have a suspicion that you may possess one or more of these genes or you may already know that you do. If it is something you want or need to know, genetic testing is possible.

At the very least, I recommend that everyone considering pregnancy get a genetic test for single-nucleotide polymorphisms, also known as SNPs. This test will show if you have any specific genetic weaknesses that may predispose you or your child to certain disorders. It can detect the genes associated with an inability to absorb folate and those associated with pathways that create genetic methylation, which I explained in Chapter 2 is strongly related to autism risk.

This is something to discuss with your physician, functional medicine specialist, or functional neurologist, who can counsel you and help you find a laboratory to have the work done. I have listed some companies that perform these tests in the Resources section. Be forewarned that genetic testing can be expensive and most likely is not covered by your insurance. Though genetic testing is not commonly done, I suspect that it will become common sometime in the not-too-distant future.

Something to consider if you are considering genetic testing is your own family history. Do you have a family history of autism, or relatives with autism-like traits? Are many of your relatives left-brain dominant? Does anxiety and antisocial behavior or schizophrenia run in your family?

True genetic mutations, such as those that cause fragile X syndrome, are unchangeable. However, many others are modifiable. It is estimated that autism involves at a minumum five to ten genes and possibly a hundred or more. Many have been identified and there are possibly more that have yet to be discovered. If genetic testing is something you want to consider, this a list of genes with a known connection to autism. Use it only as a guide, as it should not be considered a record of all of them:

Gene CDH9, CDH10

Gene MAPK3

Gene SERT (SLC6A4)

Gene CACNA1G

Gene GABRB3, GRBAA4

Gene Engrailed 2 (EN2)

Gene Reelin

Gene SLC25A 12

Gene HOXA1, HOXB1

Gene PRKCB1

Gene FOXP2

Gene MECP2

Gene UBE3A

Gene Shank3 (ProSAP2)

Gene NLGN3

Gene MET

Gene Neurexin 1

Gene CNTNAP2

Gene GSTP1

Gene PRL, PRLR, OXTR

Gene KCTD13

CHAPTER ELEVEN

The Brain Balance Program for Parents-to-Be

The principle behind getting a brain in balance is simple: Stimulate the sluggish side without ratcheting up the strength of the other side. This is exactly what we do to correct neurological problems in children at our Brain Balance Achievement Centers. At Brain Balance, we target specific areas within each hemisphere in a precise way, first focusing on areas of the brain that are most obviously challenged. In this chapter, you will learn how to do this as well.

You've taken the assessment in search of your weak areas. This chapter includes exercises you can work on to help correct an imbalance. They have been adapted for adults from some of the exercises we use with children in our Brain Balance Achievement Centers and were also featured in my book *Disconnected Kids.* You do not necessarily have to implement all these exercises, only those in which your assessment indicates you may have an imbalance. For example, if taking the assessment found that there is not a large difference in your sense of touch from one side of your body to the other, you can ignore the section on touch and tactile stimulation exercises. On the other hand, if you have an obvious difference in sensing smell or sound from one side to the other, then you should focus on the exercises involving sound, music, and smell.

These exercises are not intended to be done all at once. Begin with one area. When the imbalance between the two sides of the brain is gone or minimized, then stop doing that activity and move on to another activity until the imbalance appears to be gone. As your brain starts to get more into balance, you can begin combining your exercises.

Periodically—about once a month—you should reassess yourself and your lifestyle to make sure that the imbalance is not coming back. Remember, your goal is to be in the range of 0 or 1. Whatever factors produced the imbalance before may still exist and can possibly creep back to create an imbalance again. This is more true for an adult than it is for a child. To this end, you should practice lifestyle modifications at the same time you are doing this program. If you don't change your lifestyle, the problem will return.

DOMINANCE PROFILE

We have found that most kids are able to realign their dominance profile quite readily as they go through the Brain Balance Program. In adults, dominance is virtually impossible to change because it has become ingrained over the decades, not only neurologically but behaviorally. However, it doesn't hurt to try.

You only need to do this if you have mixed dominance, meaning you do not naturally use the hand, leg, eye, and ear all on the same side. Usually it is the foot that does not coordinate with the rest of the body. Most people with mixed dominance lean toward right dominance, unless their parents or other people in their family lean mostly left. Only work on the body part or parts that are out of phase with the rest of the body.

Hand and Foot

Wherever it is that you are out of sync, simply use the nondominant side more. For example, if using your hand is the only thing dominant on your right side, try using your left hand more for common tasks, such as brushing your teeth, reaching to pick up objects, and even writing. If it's your leg, get a ball and start kicking it around with your less dominant leg. One day use your right hand or leg, then the next day use your left.

Eye Dominance

Eye dominance is a little tricky. If your nondominant eye is on the opposite side of your brain weakness—that is, left eye and right brain—you can help strengthen that eye by wearing a patch over the dominant right eye to reduce light stimulation to the brain. Only wear one when you are doing something that could not result in injury. For example, don't use a patch when driving a car or walking down stairs. However, you can wear a patch while working on the computer or watching television. Remove the patch if you feel your eye fatiguing or getting tired. Do this exercise for longer periods of time until you work up to a few hours a day.

There is another way to work on eye dominance without reducing light stimulation to the brain. If you don't wear glasses you can use safety glasses. Here's what to do: Tape a piece of clear green acetate over the left lens to wear throughout the day. When you read, put a large piece of clear red acetate over whatever you are reading. This will do several things. First, it will block the words from the left eye. The

> **BRAIN BALANCE RESOURCES**
>
> *Disconnected Kids* offers a complete and extensive program for balancing the brain in children using sensory motor exercises. They can work just as effectively in adults. A functional neurologist can also design a program to fit your individual needs. For the best results, seek professional guidance from a functional neurologist.

red and the green will mix and it will black out the words to the left eye only, although you won't realize it.

The color green is more specific to the right brain, so using this color will stimulate the right brain with light while allowing you to only use the right eye to read. Red is specific to the left brain, so it will only come into the left eye and stimulate the left brain.

Ear Dominance

In order to hear words and syllables most effectively, most people should be right-ear dominant. If you are left-ear dominant, there are several exercises you can try:

1. Hearing your own voice only through your right ear is very powerful. Use an earplug to block sound in your left ear, and read aloud to yourself. Do this for about twenty minutes to start, and build up to an hour.

2. Listen to music with only the right ear using the same method. This is something you can do often and for a longer period of time while you are doing other things. However, it does not have the same impact as hearing your own voice.

3. An even better solution is to plug the left ear and listen to music specific to the right hemisphere. Research has demonstrated that there are defining sounds that are specific to one side or the other. I have created the only completely hemispheric-based music called Brain Balance Music for this reason. I worked with a very talented composer and musician for years to create the music, which we verified through laboratory testing. You can purchase it by going to brainbalance music.com.

STIMULATING THE OPPOSITE SIDE

When you did your assessment, you tested seven sites in which weakness in the brain is on the same side as weakness in the body. I am not going to give you specific exercises for each one. However, be aware that any time you want to work on a weakness, you will have to suppress the strong side and stimulate the weak side simultaneously. These imbalances should naturally align when the brain itself gets more balanced through your work with a functional neurologist.

Motor Exercises

If you have a left-brain weakness, you can help correct it by doing fine motor skill activities that require using both hands at the same time, such as playing the piano or typing.

If you have these muscle imbalances—and most people with a brain imbalance do—you need to increase your activity level that uses your core muscles as a whole. These are exercises such as yoga, tai chi, walking, and jogging. They can all help stimulate areas of the brain that govern gross motor muscle control and help reduce the imbalance. However, you should not start an exercise regimen without having core stability, even if you are only planning to walk. Remember, it is all about balance. Core stability is all about balancing the flexor muscles in the front of your body—your abs, girdle, and hips—and the extensor muscles in the back of your torso—your shoulders and spinal muscles. For good stability the extensor muscles should be at least 50 percent stronger than the flexors. Studies have shown that in almost everyone with chronic back pain this ratio is off or completely reversed.

Core Stability Exercises

Back muscles are our high-endurance muscles. When they are not used enough they atrophy faster than most muscles. The abdominal muscles do not atrophy nearly as fast as the back muscles. So in someone who is sitting all the time, the back muscles atrophy much faster than the stomach muscles, which reverses the ratio you need for core stability.

People frequently start a workout program doing crunches to tighten the stomach muscles. However, this only gets the ratio out of balance and creates more instability. If you start aerobic exercise on an unstable spine and core, it increases inflammation and the stress response and eventually may lead to pain and injury. Then the inevitable happens: The exercise program gets put on hold and you go back to sedentary living.

Initially I recommend doing core stability exercises three to five times a day for as long as you can manage, which will be less than a minute for each one. Repeat them until you can do each of them for a full minute five times a day.

These basic core activities should help get you ready for an exercise regimen. Work your way through all three levels. However, you can start slowly on an aerobic program after mastering both level 1 exercises.

Supine Core Stability Exercises

LEVEL 1

Lay on the floor with your hands by your side and your palms on the floor for stability and your legs a shoulder width apart. Lift your trunk off the floor as far as possible, keeping your spine straight. Hold this position still for as long as possible.

Use a stopwatch or count to yourself as you hold the position. If you start to shake, bend, or collapse, stop and take note of the time. An adult with core stability should be able to hold this position for at least sixty seconds. Work up to it before moving to level 2.

LEVEL 2

Repeat the Supine Core Stability level 1 exercise, relax for ten seconds, then do this exercise:

Lie flat on the floor with your knees touching. Cross your arms over your chest. Again, lift your trunk off the floor as far as possible and hold the same spine-straight position. Count the number of seconds you can hold the position. You should be able to hold this position for sixty seconds. Once you are able to do this five days straight, move on to level 3. At this point you can now start to do aerobic exercise.

LEVEL 3

Assume the same position and lift your torso as in level 1. When you're in position, lift one leg off of the floor about an inch. Count how long you can hold this position. Switch legs and repeat. Record your time for each leg and work up until you can hold each leg for one minute.

Prone Bridge Core Stability Exercise

LEVEL 1

Lie facedown on the floor with your arms and legs straight out. Raise your head and one arm off the floor about six to eight inches, keeping the arm perfectly straight. Hold this position for as long as possible, for up to fifteen seconds. Count how long you can hold this position and record it below. Repeat this with all four limbs separately until you can maintain each for fifteen seconds, a total of one minute. When you master this, move to level 2.

LEVEL 2

Assume the same position. This time you will lift your head, one arm, and the opposite leg off the floor. Count how long you can keep the position. Relax for ten seconds. Switch arm and leg and repeat. Work up until you can do thirty seconds on each side.

LEVEL 3

Assume the same starting position. This time lift all four limbs off the floor simultaneously, as in a Superman position. Keep track of how long you can keep the position until you can do it for a full minute.

Sensory Stimulation

Doing sensory stimulation exercises is one of the most effective ways to help balance the brain. If you have a significant imbalance, your best bet is to find a functional neurologist who can give you a full evaluation and can put together an individual program that includes motor activities, sensory stimulation, and cognitive-based exercises. However, these are some basic exercises you can try.

Use the stimulation on one side only until you can sense the two sides are more balanced.

Light
Block the eye on the same side as your brain imbalance with a patch and shine a light in the corner of the opposite eye. Use colored light: red, orange, and yellow if you are blocking the left eye, and blue, indigo, and violet if you are blocking the right eye.

Sound
Use an earplug to block sound in the ear on the same side as your brain imbalance, so you are listening through the opposite ear. Listen to Brain Balance Music designated for either a left- or right-brain deficiency.

Tactile
Tactile stimulation can involve using light touch, massage, a vibrating massager, or a brush on the arm and/or leg opposite the side of the brain you want to stimulate.

Smell
Smell is the only sensation that goes to the same side of the brain. Pleasant smells can be used to stimulate the left side of the brain and

strong, pungent, or unpleasant smells can be used to stimulate the right hemisphere.

Temperature

Warm and cold both work, but the best thing is to use the sensation that you can feel the least. Use it on the hand, foot, arm, and leg on the opposite side of the hemisphere you want stimulate.

Balance

To help balance the inner ear, do the same head-turning activity you tested in Chapter 7, but concentrate only on the side that was harder to focus and didn't cause dizziness. Do it ten times. If you are turning to the right, stick your right hand out and look at your right thumb. If you turn your head left, put the left hand and thumb up. Do this until all symptoms are gone.

MENTAL ACTIVITIES

Mental activities that are done regularly to stimulate one side of the brain or the other can be fun and very effective.

Right-Hemisphere Activities

- Sing

- Rhyme

- Write a song

- Try to make up a rap song

- Try creative writing

- Perform pantomime

- Do mazes

- Draw maps

- Try to copy someone's unusual body position

- Draw complex geometric shapes

- Draw or look at something from an unusual perspective

- Draw or paint something upside down and backward, especially with your nondominant hand

- Try to recognize familiar faces by looking at them upside down

Left-Hemisphere Activities

- Recite sequences of numbers

- Count backward

- Read aloud

- Spell

- Identify objects from an unusual viewpoint

- Ask or have someone choose a letter or a category and then name as many things that start with that letter or are in that category as you can in one minute; keep track and try to improve

- Do calculations in your head

PART 3

EARLY DETECTION

CHAPTER TWELVE

Searching for Early Signs of Autism—and What You Can Do

Most doctors say that autism can't be detected until around age two and a half, three, or even four when symptoms start to become obvious: Your child doesn't make eye contact. He doesn't play with others. He's not talking. He doesn't want to be cuddled. You notice that he's just different from other kids.

This isn't my experience, nor is it the observation of the few researchers who are investigating to find out if autism can be spotted at an earlier age, when more can be done to stop and correct the problem. From our perspective, it appears there is much we can do to find autism at a much earlier age. In fact, researchers have found it is possible to see signs of autism as early as six months of age. This was first demonstrated more than ten years ago by research scientists Osnat and Philip Teitelbaum, both PhDs at the University of Florida, who found that they could predict whether babies would eventually be diagnosed with autism simply by watching them interact with their world.

The babies who ended up with autism had trouble rolling over and couldn't sit up without assistance at six months old—skills that should have been mastered by this age. It was a clear sign that they had not yet let go of their primitive reflexes, which were preventing their more mature postural reflexes from taking over. It meant their motor skills were not developing on target. They were missing important developmental milestones.

Most recently, a study at Kennedy Krieger Institute in Baltimore came to a similar conclusion. Research on six-month-olds at high risk for autism found that 90 percent of those who were eventually diagnosed with autism exhibited head lag on a simple "pull to sit" task.

John N. Constantino, MD, of Washington University, found uncoordinated motor movement is so common in autism that it should be one of the core distinguishing symptoms used to diagnose autism early on. One main commonality Dr. Constantino found was that children with autism display motor problems, while their normally developing siblings do not.

Many pediatricians tell parents that milestones are not all that important, that children tend to develop at their own pace. This is true only to a small degree. Milestones are *very* important. They are signs that the brain is developing normally. From the time your baby is born you should pay close attention to milestones. Is she turning over, sitting up, crawling, walking, and talking on schedule? Does she have the natural movements expected in infancy and early childhood? Is she keeping up with other children her own age?

If an infant or toddler is having a problem with motor progression, then they most likely are developing other issues, including digestive problems and immune and hormone imbalances, which need to be addressed as well. The best and earliest way to spot if something might be amiss is by tracking your baby's primitive reflexes.

PRIMITIVE REFLEXES AND
WHY THEY'RE SO IMPORTANT

Primitive reflexes are the basic necessities a newborn needs for survival. A baby starts life with little brain—not much more than a brain stem—and very little muscle tone. This is why primitive reflexes exist. They don't require any thought. They give baby the instinct to breathe, to feed when hungry, to squirm and cry when uncomfortable, to coo when cuddled. Each reflex plays a part in the necessary growth of the fetus and infant, and paves the way for each stage of early development. At three to five months, primitive reflexes herald one of the first and most important milestones—rolling over.

Primitive reflexes are important even before birth. They develop while in the womb to help a baby make the trip through the birth canal. If they are faulty, a child may have a difficult birth. Breech is a sign that these reflexes may not be fully activated and symmetrical. Caesarean birth means the baby misses his first opportunity to use them. In these cases, an observant obstetrician and pediatrician could be looking for signs of a brain imbalance right from birth. Unfortunately, this rarely happens.

Muscle movement is key to healthy brain development. It prompts genes to build the brain and grow the neurons and connections that advance a newborn from one milestone to the next. As the brain gets larger and more connections form, higher levels of brain function are ready to come into play. These new connections eventually inhibit the more primitive baby movements, setting the stage for more complex motor reflexes, known as postural reflexes, to emerge. At around four to six months of age these primitive reflexes, for the most part, are no longer needed and should be replaced by postural reflexes. Postural reflexes allow both sides of the body to move in perfect synchrony and cooperation. They are key to being able to stand on two feet and walk smoothly, coordinatedly, and effortlessly. Once a baby develops this ability on or near age one, he can lift himself up and walk.

Primitive reflexes never totally disappear, but rather than being independent, they come under control of the brain. The brain breaks them up into bits and pieces and reassembles them to create smooth coordination of complex movements.

If a child does not move enough to stimulate genes to build the brain, the reflexes will not become inhibited. If primitive reflexes aren't inhibited, the brain doesn't develop in an orderly fashion. Anything that disrupts the order adds to the risk of the brain becoming out of sync. This is because the brain can't "leapfrog" from one stage to the other. If it misses the construction of an important skill, it can't go back and activate it later on. Once the window of opportunity is missed, it is lost. This is how an imbalance begins. Depending on how the imbalance unfolds—if step after step is missed—it could result in autism or any other neurological condition. However, it is possible to correct or at least improve the trajectory of errant development. The earlier we can intervene, though, the easier it is to restore balance and proper development. Looking at your child's primitive reflexes and postural reflex development are important diagnostic tools. Most primitive reflexes should be active at birth, and most should fade out by around six months of age and eventually be replaced by consciously controlled or directed movements.

It is actually possible that *you* may still have some active primitive reflexes, especially if you lean toward an autistic phenotype or have multiple health issues or a brain imbalance. I have seen this in many adults. They have their own stories of childhood struggles that today we can say would be linked to a developmental disorder. I have found active primitive reflexes in every autistic adult whom I have examined, and I would wager that most if not all autistic adults have active primitive reflexes. The good news is that it takes relatively little effort to inhibit these reflexes in babies, older children, and even adults. This is a very helpful intervention, because it helps decrease the imbalance between the two sides of the brain.

WHAT PEDIATRICIANS DON'T SEE

Primitive reflexes are a well-documented crucial stage of childhood development. In my experience, however, pediatricians rarely check them in a physical exam, and almost never check for them after the age of six months, when they are supposed to disappear. Even most pediatric neurologists tell me they don't consider checking for primitive reflexes in a child after six months, even though pediatric neurology textbooks say their presence after this age is suggestive of brain

---

HEAD LAG TEST

This simple test is easy to perform.

Lay your baby on his back on the floor or changing table. Grab both wrists or allow him to grab ahold of your hands. Slowly and gently pull your child up by the arms.

It is normal for the head to lag behind the body until around three months of age. It is also normal for the child to not pull forward or try to assist you in any way. However, by around four to six months your child should be able to lift his head and slightly flex it forward, or even pull slightly with his arms. When doing the test, it is important that you pull slowly; if you pull too fast it will cause any child to have head lag.

When the test is done correctly, head lag at six months can be a strong warning sign of autism.

---

injury. Clearly, primitive reflexes are not high on the radar of traditional pediatric medicine, despite the fact that they can provide a clear indication that development is on track.

If a baby is delayed in acquiring a motor milestone, it is most likely because his primitive reflexes are still active. This can happen at any time within the first year, but especially within the first six months. Ask your pediatrician to check for them. If the doctor says she doesn't do that kind of test because primitive reflexes are not important, then I suggest you consider getting a new doctor. Whether or not your doctor does the exam, you should also do it yourself. It is important to know if primitive reflexes are functioning in newborns and that they are starting to fade away at around four months.

Not too long ago, I was giving a lecture to health professionals on how to check for primitive reflexes and was approached afterward by one of the doctors. She said she had a child with a developmental disability, despite doing everything possible to have a healthy pregnancy. She was now seven weeks pregnant and worried that the baby might follow the same course. As a result of the lecture she realized she still had most of her primitive reflexes.

"It explains a lot about all the struggles I had growing up," she told me. "What does it mean to this baby if my first child has a disability?" I said there was no way of knowing, but we could possibly help by inhibiting her primitive reflexes. With little effort, we eliminated the reflexes in her using the same exercises I am offering you. I also did some interventions to help balance her brain, which included fixing her gut and balancing her immune response. She went on to have an incident-free delivery, which in every way was significantly better than her first pregnancy, and a child with normal primitive reflexes and no apparent developmental delay.

Testing Your Baby's Primitive Reflexes

There are more than thirty different primitive reflexes, but there are only a few that are consistently associated with autism. These are the ones to watch. All of these primitive reflexes should be present at birth. All should be gone by the age of four to six months, except for the plantar reflex, which should be gone by age one. You should check your newborn to make sure all the reflexes are present, then start checking at around four months to make sure they are starting to disappear.

You may notice that some of these reflexes may be more persistent on one side of the body. This may signify that one side of the brain is developing normally but not the other. This could also mean a nerve injury or possibly even a brain injury. Have your child checked out by a doctor as soon as possible. Most likely, the doctor will find there is no nerve damage or brain injury, but this is something that should not be ignored. Motor milestones do matter, and it's important to monitor them to make sure they're on track.

If you have toddlers or older children, you should check them to make sure their primitive reflexes are gone. If they aren't, I offer exercises that should get rid of them. As a prospective parent, you should also check to make sure you don't still have primitive reflexes. If you do, the same exercises I offer for older children should be effective for you as well.

If you have a child over age three who is showing signs of a development delay, the program in my first book, *Disconnected Kids*, should be helpful. In my experience, early intervention means these problems can be reversed most of the time.

Remediating Primitive Reflexes

If you spot primitive reflexes still in evidence at six months, remedial action is straightforward. The way to get rid of them is to use them.

EARLY WARNING SIGNS OF A
DEVELOPMENTAL DELAY

- The first sign is extremely low muscle tone. This is often first observed when a child cannot properly suckle at the breast.
- Your child should roll over easily and to both sides at around three to five months. If this does not happen on time, it is a sign that a brain imbalance might already be present.
- Your child exhibits digestive problems: He has problems with bowel movements and constipation and/or has a problem with formula, especially dairy and possibly soy-based products. He may also have reflux.
- Your baby or toddler develops eczema or chronic ear infections.
- Your child should crawl and then creep in proper fashion. Crawling is a commando type of crawl—arm over arm propulsion forward with the trunk on the ground and legs dragging. Crawling is usually seen at seven to nine months. Creeping is crawling on hands and knees with the trunk off of the ground. Creeping is usually seen at eight to twelve months. Anything else, such as scooting on the bottom, rolling instead of crawling, or dragging one leg are red flags.
- Skipping crawling altogether or walking too early are important warning signs. A child should be walking at eleven to thirteen months. In my experience, walking earlier or later than this age signifies a developmental delay. This is counter to conventional thinking that puts the walking window at eleven to sixteen months. If a child was born premature, the window should be extended; for example, if the child was one month premature, then consider walking at fourteen months to be within a normal range.

Lack of use is the reason they persist. In an infant, you remediate the same way you test. Just bring on the reflex as described below and repeat it until it fades. Repeat it in rapid succession five to ten times until you can no longer bring it on. At this point stop, and a few hours later go back and do it again. Do this several times a day until the reflex is permanently gone.

Primitive Reflex: Root and Suck

The first primitive reflex a baby adapts at birth is the rooting and sucking reflex, which is the breastfeeding reflex. When the skin around the baby's mouth is stimulated by touch, it initiates the rooting reflex. Baby will open his mouth and turn toward the stimulus. The sucking reflex allows baby to latch onto the breast and feed. Many new mothers have difficulty breastfeeding because this reflex is not formed. This can lead to failure to thrive because the baby can't suck on a nipple to feed. Likewise, this reflex should fade when your baby no longer needs it.

Your baby should have a strong coordinated suck with good stripping action. You can test it using a pacifier. To test for the root reflex, use your finger or a small paintbrush to gently stroke your baby's cheek about a half inch from the corners of the lips. Make the stroking motion toward the mouth. Baby should open his mouth toward the stimulus and turn his head to latch onto the pacifier. There should be resistance when pulling out the pacifier.

You'll know if the sucking reflex is present if your baby turns and opens his mouth toward the stimulation or if there is an involuntary twitch of the muscles around the mouth, lips, or face. This should be gone at four to six months. More often, however, the problem is that this doesn't go away. To help ensure that it does, do this exercise several times a day until it is gone. If it doesn't go away, the child may not want to give up a pacifier, which could lead to a problem with speech articulation or a lisp.

Primitive Reflex: Moro

The Moro reflex is the infant startle response. When a child hears a loud noise, she should be startled and jump , which is usually followed by wailing. This reflex is also stimulated if the child feels like she is falling backward. The reflex is there to help the child hold on to the mother and not fall or be dropped.

To test the reflex, place your child in the fetal position and clap your hands loudly or make a loud banging sound. She should react with a startle. The arms and legs will usually fly open, the head and back will lean backward, and the baby will take a deep breath before going back to the fetal position. This happens very quickly. The noise will kick in the stress response, which you will see as flushing and possibly crying.

Another way to elicit the Moro reflex is by lifting your baby's head and shoulders just a few inches off a mat. The arms will automatically flex to the chest in the natural fetal position. Gently let the head and shoulders drop back to the mat.

Another way to test this is to hold your baby tightly, supporting her head with your hand, facing you. Quickly thrust your upper body forward a few inches, which will make your baby feel like she's falling. Your baby's arms should fully or partially fly out from her body and her hands may fly open as well. Return your body to the original position.

An asymmetric Moro response, in which only one arm flies out and the hand stays open, may indicate a brain imbalance. A Moro that doesn't go away may be seen as a developmental delay. An absent or incomplete Moro may indicate brain immaturity or possibly injury.

The Moro is usually gone by four to five months of age. Persistence of the Moro reflex beyond this time is a sign of a developmental delay. If it is still present after four to six months, stimulate the reflex over and over until it is gone. For older children and adults, you can get rid of this reflex through an exercise known as the Starfish. For a child, the exercise requires close supervision by a parent or adult.

STARFISH

Have your child sit in a chair in a fetal position as shown in the illustration, with the right wrist crossed over the left and the right ankle crossed over the left ankle. Fists should be closed. Ask your child to inhale and make like a starfish by swinging his arms up and out and thrusting his legs out while extending the head back and opening the hands. Have him hold this position for five to seven seconds while holding his breath. Then tell him to exhale and return to the same position, crossing the left wrist and ankle over the right wrist and ankle. Repeat this again until he is back to the original position. Do this six times in a row a few times a day until the reflex is inhibited fully.

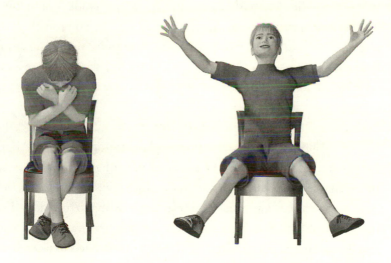

Primitive Reflex: Galant

This is the reflex that helps a newborn move through the birth canal. It also helps a newborn roll over. The reflex can be activated by stroking the skin along one side of the spine in a downward motion. This will cause your baby to flex the muscles on the same side by bending forward into a "C" shape.

For an older baby, place the baby on the floor on all fours, then stroke on the side of the spine about an inch below the shoulder blade. Make a quick, gentle, but firm stroke from top to bottom a few inches along the side of the spine to stimulate the skin.

This reflex should be gone by four months of age. If it is not gone, stroke the spine repeatedly to fatigue or until the reflex stops reacting. Do the following exercise for older children or an adult.

ANGELS IN THE SNOW

Have your child lie face up on a mat or flat surface with his legs extended and arms at the side. Have your child breathe in and simultaneously spread his legs outward and raise his arms out along the floor and overhead, with the hands touching. The hands should touch at the same time the legs are fully extended. Exhale and return to the original position. The key is to get your child to move all four limbs slowly at the same time. Do this five times a day several times a day until you no longer elicit the reflex.

Primitive Reflex: The Palmar Grasp

From the moment of birth, an infant will grasp your finger and hang on for dear life when you stroke the palm of her hand. This is normal for the first few months. However, if the child continues to reflexively grasp on to a finger, it prevents her from progressing to individualized finger movements. This can affect a child's ability to grasp a crayon or pencil properly. Many children with this problem need to use a fatter pen or pencil and will have writing difficulty. The grasp reflex is usually gone by four to six months of age for the hands and six to twelve months for the toes. The reflex is gradually replaced by the voluntary activity of reaching and grasping with the hand.

Test one hand at a time using your finger or the handle of a small brush. Gently stroke from the outside of the hand to the middle of the palm, moving quickly from all four directions. The reflex will make your baby's hand close automatically. Open the hand and repeat it two more times. Test both hands. Repeat until it goes away.

PLANTAR REFLEX

Also known as the Babinski sign, the plantar reflex is slightly different from the grasp response. To test for this reflex, hold your baby's foot by the ankle and use your fingernail or a small brush to gently stroke the outside bottom of your baby's foot, from the bottom of the heel, along the outside of the foot toward the toes, but don't touch the ball of the foot. Do one foot at a time. Use the same amount of pressure you would to tickle the foot. If the reflex is present, the baby's big toe will go up and all toes should also go up and fan out.

This reflex should be present at birth and be gone by the end of twelve months. When it is gone, all the toes will go down when the foot is stroked in the same way. It is a great way to assess the development of both sides of the brain. If all the toes of one foot go up and the toes of the other foot go down, it is an indication of a developmental brain delay on the same side where the toes go down. If there is a delay on one foot more than the other, it can be an early sign of a developmental delay or the beginning of a brain imbalance.

To eliminate the reflex, repeat the exercise over and over to fatigue. Do this several times a day until the reflex is permanently gone.

In an older child or adult, have them pick something up off the floor using their toes, preferably the foot opposite of hemisphere weakness. This will increase the activation of that hemisphere. It is more difficult than using your hand, and therefore it is more effective at exercising the opposite side of the brain.

Primitive Reflex: Asymmetric Tonic Neck

The asymmetric tonic neck reflex is one of the most common reflexes to persist in autism. This reflex is believed important for movement through the birth canal. It also helps a baby to corkscrew her body to roll over, and it helps her start to creep and crawl. If a child doesn't crawl properly on time, or at all, it is often because this reflex is still present.

This reflex should be present until three to four months of age, at which time the baby should be able to overcome the reflex and move out of this posture. If the infant cannot move out of or overcome the reflex or if the reflex persists beyond six months of age, it is abnormal and should be checked out by a doctor. It can be a sign of a brain injury, but it is more likely a sign of a developmental delay or a brain imbalance, especially if it happens on one side and not the other. If it persists on one side more than the other, it signifies an imbalance in the development of the nervous system, and the child will usually only be able to roll to one side.

To elicit the response, make sure your baby or child is lying calmly on his back on the floor. Gently move his head to one side and hold it for ten seconds to initiate movement of the legs and arms. It should take a few seconds for the reflex to start, and it will get more pronounced as you hold the head still. Your baby should respond by straightening and extending her arm and leg on the side of the head turn and then flexing the arm and leg on the opposite side, like a fencing position. Repeat on the other side. Make sure you are moving the head gently and naturally, as you do not want to cause a neck injury.

This reflex movement is normal and expected for the first four to six months. If this movement occurs when turning the head in a child who is older, consider it abnormal. Just keep repeating this activity every day several times a day until this reflex is gone. In children age four and older, and in adults, you can inhibit this reflex with the Fencer exercise.

FENCER EXERCISE

This one may take some practice to get right, but be patient. Have your child sit in a chair and turn his head from side to side, or to the

WHY KIDS DON'T NEED COMPUTERS

Most of the genes that build functional connections in the brain are experience-dependent genes, meaning they do not respond to chemical activation. Instead, they engage as a result of physical movement and sensory stimulation. This is why it is so important for kids to go out and play instead of sit in a chair and play games on a computer or other electronic device.

Movement is one of the most important stimuli for healthy brain development. Genes that build functional connections in the brain are turned on when a child physically interacts with the world. This is why computers, as amazing as they may seem to be for brain development, are in fact just the opposite. As our environment becomes more and more dependent on computers, our children are becoming less and less involved in physical exploration. Rather than move around and examine the world, their bodies sit still while they entertain themselves with technology.

Boys tend to be affected the most. They are more attracted to television, computers, and video games—a left-brain trait. Studies show that children who are the most technically capable are also the most socially impaired, one of the main traits of autism. Socialization is a right-brain learning skill. Studies also show that computers and digital information increase the activity in the left brain but decrease activation in the right brain. All of this together makes boys more susceptible to a right-brain deficiency. It's another explanation for the high rate of autism we are seeing in boys.

one side that still elicits the reflex. As your child is turning his head, have him extend the foot and arm of the same side outward from the body and look at his hand. The opposite hand should also open, the arm should flex, and the other leg should bend. Have the child return to the starting position, and repeat until the reflex fatigues. Repeat three times in a row. Once this exercise is done correctly, three times very slowly on each side is usually enough.

Two Postural Reflexes to Look For

As I mentioned earlier, in a normally developing baby, primitive reflexes fade, allowing postural reflexes to take their place. However, this is not what we typically see in children with autism. For them, postural reflexes are almost always delayed, due to persistence of primitive reflexes that are no longer needed.

When a baby sits up without falling over, or extends his arms when falling forward, he is using his postural reflexes. It is a sign that the baby has inhibited his primitive reflexes and is maturing and developing appropriately. Here are two key postural reflexes to check for in your growing baby. Unlike primitive reflexes, which should fade away by a certain age, these reflexes should develop by a certain age—and remain in place for life.

Postural Reflex: Lateral Propping

Lateral propping, or propping oneself up by extending an arm to catch oneself when falling to the side, is essential for a baby to sit up on his own, and it should develop at five to seven months of age. To test this reflex, kneel behind your baby or child while he is sitting up, and place your hands on the side of each shoulder. Gently push him to one side with a quick shove, so that he loses balance. Keep your other hand in position to prevent the fall. The shove should make him react by throwing out his arms to protect himself.

If the reflex is present on one side and not the other, this is known as asymmetric lateral propping and can be an early sign of a developmental imbalance.

Postural Reflex: Parachute (Anterior Propping)

When a child runs and falls, his arms should automatically fly forward to catch himself and protect his face from hitting the ground. Likewise, if you hold a baby or a small child and pretend to dump him forward facedown and upside down while you hold him, his arms should automatically fly overhead to protect himself. This reflex, called the parachute reflex or anterior propping reflex, is the last of the postural reflexes to develop. It usually appears at eight to nine months of age. Prior to developing this reflex, a baby will actually bring the arms back to their body or hold on to your hands.

READY FOR DEPARTURE

Once the primitive reflexes are gone and the postural reflexes are present, you can stop doing the exercises. If the primitive reflexes I just reviewed are present after the specified age or the postural reflexes are delayed, it is likely that other primitive reflexes are still active as well. Once your doctor has ruled out any brain injury or nerve damage, I recommend that you seek out a professional knowledgeable about brain balance or functional neurology who can do a full assessment and instruct you on other exercises. We might not be able to diagnose autism at age six or eight months, but these early milestones are strong indicators as to whether a baby's brain is developing as it should.

Getting Back Our "Right Mind"

I believe that our environment and society as a whole have been gradually causing more and more of a shift toward left-brain dominance. It is reflected in the way we think, the way we behave, in the way our education system is run, and in the routines that dominate our daily lives. This shift has been slow and gradual over the past twenty years and is affecting our young ones at an earlier and earlier age.

Almost from the moment of birth, children are bombarded with a sensory overload from television, computers, video games, and other electronic devices, as well as "educational activities" such as games and flash cards designed to help them learn while practically still in the cradle. It's important to understand that the brains of babies and small children are not miniature versions of an adult brain. Their brains haven't even been built yet! Yet modern society, motivated largely by marketers targeting well-meaning parents, is attempting to get them to act and perform like adults at an earlier age. Parents are urged to treat small children like they are little adults and made to feel guilty if they don't follow the herd. This is an unfortunate trend.

Adulthood is largely about familiarity and routine, but childhood is about novelty and new experiences—in fact, babies and young children are experiencing *everything* for the first time. A newborn greets the world as wondrous and new and he can't wait to explore it. The brain is designed to lead this adventure as it is being built. Prenatal and early childhood development are primarily a time of right-brain dominance and development. It's the time to get the big-picture (right-brain) view of the world. The details (left brain) will come naturally in due time. It is not necessary or beneficial—indeed, it can be detrimental—to intervene with activities, lessons, and experiences presented on an artificial timeline. To put it bluntly, no parent is doing his child a favor by trying to teach her to read before she is potty trained.

Childhood is more about right-brain development than left-brain development. Our society has been increasingly limiting right-brain development in childhood, and I believe this is increasingly turning them into left-brain dominant adults. Nature has created a delicate balance between the two sides of the brain, but we are disrupting this balance and creating an environmental shift that is threatening the developmental health of our children.

The brain is intended to develop in an orderly fashion. There is a constant battle for control and dominance of one hemisphere over the other. This competition is partly what makes the human brain unique, but it also creates vulnerability. If we develop an imbalance on one side that is too dominant, it will continue to suppress the opposite side to the point where it cannot correct itself, so that it will become more imbalanced as time goes on. This decreases a person's ability to function well in the world.

When this dominance persists in childhood it often becomes exaggerated in adulthood. And it perpetuates itself as parents pass these traits on to their children. I believe it is at the root of the rise in autism rates as well as many other illnesses, such as obesity and diabetes, that are increasing at alarming rates around the world. Environmental and behavioral factors are combining with genetic predisposition, and then being passed on to the next generation.

I am not alone in this view. Iain McGilchrist, MD, a psychiatrist and the author of *The Master and His Emissary: The Divided Brain and the Making of the Western World*, describes brilliantly how our modern society is drifting more and more toward left-brain type of thinking, which he believes is creating an explosion of neurological disease and unhappy people. "Nowadays we live in a world that is paradoxical," he writes. "We pursue happiness and it leads to resentment and it leads to unhappiness and it leads, in fact, to an explosion of mental illness . . . In our modern world we've developed something that looks an awful lot like the left hemisphere's world. We prioritize the virtual over the real, the technical becomes important, and bureaucracy flourishes. The picture, however, is fragmented. There's a lot of uniqueness. The how is subsumed in want and the need for control leads to a paranoia in society that we need to govern and control everything."

I believe that children with autism are the proverbial "canary in the coal mine." They are the most vulnerable among us, and in them I believe we see an exaggerated version of our society as a whole.

I believe that this epidemic of autism and other similar neurological disorders is a product of our environment, and that it is a powerful wake-up call to all of us, telling us that we need to care more for our planet and one another. We need to explore the world around us and reconnect, person to person. We need to reestablish a balance between good old-fashioned ideals, thoughts, and behaviors with the convenience of modern technology and information. In short, we need to get back in balance.

Albert Einstein, who started out life with a learning disability, once said, "The intuitive mind is a sacred gift and the rational mind is a faithful servant. We have created a society that honors the servant and has forgotten the gift." He said that nearly one hundred years ago. I can only imagine what he'd be saying if he was living in our twenty-first-century world.

While we cannot control every risk factor we face, or change the modern world in an instant, we can better understand the factors around us and make meaningful changes that will affect our own health as well as the health of our children, our communities, and the planet.

It's time to get back in balance. It's my hope that this book will play a part in making that happen.

ACKNOWLEDGMENTS

First, I want to dedicate this book to my wife, Carolyn, and my children, Robert, Ellis, and Ty. Their sacrifice has allowed me to pursue these important questions and help others. Without their love and support, nothing is possible.

Second, I want to thank all of my family members for their love and support throughout the years. I want to thank all of my Brain Balance familly for their dedication to children and families in need, especially my partner Bill Fowler for his support, hard work, and inspiration. I also want to thank my functional neurology friends and research companions, especially Gerry Leisman. I would also like to thank Debora Yost for her help and support; it has been one of my greatest pleasures to work with you on all three of my books. Also, of course, my agent, Carol Mann, and everyone at the Carol Mann Agency for their support and hard work. Lastly, I want to thank my editor, Marian Lizzi, for her helpful advice and her support for my books, and her assistant, Lauren Becker, for her attention to the details. Also a special thanks to all the scientists and researchers I spoke with while doing research for this book. Their information is invaluable.

RESOURCES

If you would like to see Dr. Melillo in person to treat you or your child or have him design a personalized preconception program for you, you can set that up at www.drrobertmelillo.com.

To find a local specialist in functional medicine or functional neurology nearest to you, visit www.acfnsite.org or www.acnb.org and click on "Doctor Locator."

To find out more about functional neurology, contact the International Association of Functional Neurology and Rehabilitation (IAFNR) at the F. R. Carrick Research Institute, www.frcarrickresearchinstitute.org. Or call 516-301-5305.

For more information on autism or to make a donation for research, contact Children's Autism Hope Project at www.childrensautismhopeproject.org or call 480-926-1115.

For more information on Dr. Melillo's course and other functional neurology classes, contact the Carrick Institute for Graduate Studies at www .carrickinstitute.org or contact the registar at 321-868-6464.

For information about Brain Balance Achievement Centers or to find a center near you, go to www.brainbalancecenters.com. For information about a franchise, contact the corporate office at 201-984-3742.

To find out more information about Brain Balance Music or to purchase a CD, visit www.i-waveonline.com or brainbalancemusic.com.

For more information on gluten sensitivity or to purchase a gluten sensitivity test, contact Cyrex Laboratories at www.cyrexlabs.com or call 602-759-1245. For more information on the nutritional supplements mentioned in this book, contact Apex Energetics at www.apexenergetics.com. To place an order visit www.orderapex.com or call 800-736-4381.

For more information on hormone testing, contact Metametrix Inc. Clinical Laboratory at www.metametrix.com or call 800-221-4640.

For more information on food sensitivity testing, contact ALCAT Worldwide at www.alcat.com or call 800-872-5229.

For information on preconception vitamins, contact Kirkman Labs at 800-254-8282.

SELECTED REFERENCES

American Psychiatric Association. 1994. *Diagnostic and Statistical Manual of Mental Disorders, 4th Edition (DSM-IV)*. American Psychiatric Publishing.

Anderson JS et al. 2011. Decreased interhemispheric function connectivity in autism. *Cerebral Cortex*. May;21(5):1134–46.

Autism Now: Dr. Richard Grinker Extended Interview. 2011. *PBS News Hour*. April 11; online.

Axelrad DA et al. 2004. Dose-response relationship of prenatal mercury exposure and IQ: an integrative analysis of epidemiologic data. *Environmental Health Perspectives*. April;115(4):609–15.

Bauman ML, Kemper TL. 2008. *The Neuropathology of the Autism Spectrum Disorders: What Have We Learned?* Novartis Foundation Symposium 251. John Wiley & Sons.

Bellinger DC. 2011. A Strategy for Comparing the Contributions of Environmental Chemicals and Other Risk Factors to Children's Neurodevelopment. *Environmental Health Perspectives*. Dec. 19. Epub.

Bellinger DC. 2004. Lead. *Pediatrics*. April;113(4 Suppl):101–22.

Bouchard MF et al. 2011. Prenatal exposure to organophosphate pesticides and IQ in 7-year-old children. *Environmental Health Perspectives.* Aug.; 119(8):1189–95.

Brugha TS et al. 2011. Epidemiology of autism spectrum disorders in adults in the community in England. *Archives of General Psychiatry.* May; 68(5):459–65.

Buss C et al. 2012. The role of stress in brain development: the gestational environment's long-term effects on the brain. *Cerebum.* April 25.

Constantino JN et al. 2010. Sibling recurrence and the genetic epidemiology of autism. *The American Journal of Psychiatry.* Nov.;167(11):1349–56.

Daniels JL. 2006. Autism and the environment. *Environmental Health Perspectives.* July;114(7):A396.

Davis EP, Sandman CA. 2010. The timing of prenatal exposure to maternal cortisol and psychosocial stress is associated with human infant cognitive development. *Child Development.* Jan.–Feb.;81:131–48.

Dinstein I et al. 2011. Disrupted neural synchronization in toddlers with autism. *Neuron.* June 23;70(6):1218–25.

Durkin MS et al. 2008. Advanced parental age and the risk of autism spectrum disorder. *American Journal of Epidemiology.* Dec. 1;168(11): 1268–76.

Emck C et al. 2012. Psychiatric symptoms in children with gross motor problems. *Adaptive Physical Activity Quarterly.* April;29(2):161–78.

Eskenzai B et al. 2008. Pesticide toxicity and the developing brain. *Basic & Clinical Pharmacology & Toxicology.* Feb.;102(2):228–36.

Field T, Diego M, Hernandez-Reif M. 2006. Prenatal depression effects on the fetus and newborn: a review. *Infant Behavior and Development.* July;29(3):445–55.

Fighting Autism, www.fightingautism.org.

Folstein S, Rutter M. 1977. Infantile autism: a genetic study of 21 twin pairs. *Journal of Child Psychology and Psychiatry.* Sept.;18(4):297–321.

Fraser A et al. 2010. Association of maternal weight gain in pregnancy with offspring obesity and metabolic and vascular traits in childhood. *Circulation.* June 15;121(23): 2557–64.

Golzio C et al. 2012. KCTD13 is a major driver of mirrored neuroanatomical phenotypes of the 16p11.2 copy number variant. *Nature.* May 16;485 (7398):363–67.

Grafodatskaya D et al. 2010. Autism spectrum disorders and epigenetics. *Journal of the American Academy of Child and Adolescent Psychiatry.* Aug.;49(8):794–809.

Grandjean P, Landrigan PJ. 2006. Developmental neurotoxicity of industrial chemicals. *Lancet.* Dec. 16;368(9553):2167–78.

Hallmayer J et al. 2011. Genetic heritability and shared environmental factors among twin pairs with autism. *Archives of General Psychiatry.* Nov.; 68(11):1095–102.

Halperin JM, Healey DM. 2011. The influence of environmental enrichment, cognitive enhancement, and physical exercise on brain development. *Neuroscience Behavioral Review.* Jan.;35(3):621–34.

Herbstman JB et al. 2010. Prenatal exposure to PBDEs and neurodevelopment. *Environmental Health Perspectives.* May;118(5):712–19.

Hertz-Picciotto I, Delwiche L. 2009. The rise in autism and the role of age at diagnosis. *Epidemiology.* Jan.;20(1):84–90.

Hilton CL et al. 2012. Motor impairment in sibling pairs concordant and discordant for autism spectrum disorders. *Autism: The International Journal of Research and Practice.* Jan. 18. Epub.

Joblanka E, Lamb MJ. 2006. *Evolution in Four Dimensions: Genetic, Epigenetic, Behavioral, and Symbolic Variation in the History of Life (Life and Mind: Philosophical Issues in Biology and Psychology).* MIT Press.

Johnson CP. 2007. Identification and evaluation of children with autism spectrum disorders. *Pediatrics.* Nov;120(5):1183–215.

Just MA et al. 2012. Autism as a neural systems disorder: a theory of frontal-posterior underconnectivity. *Neuroscience Behavioral Reviews.* April; 36(4):1292–313.

Just MA et al. 2007. Functional and anatomical cortical underconnectivity in autism: evidence from an fMRI study of an executive function task and corpus callosum morphometry. *Cerebral Cortex.* April;17(4): 951–61.

Kaati G, Bygren LD, Edvinsson S. 2002. Cardiovascular and diabetes mortality determined by nutrition during parents' and grandparents' slow growth period. *European Journal of Human Genetics.* Nov.;10(11): 682–88.

Kalkbrenner AE et al. 2012. Maternal smoking during pregnancy and the prevalence of autism spectrum disorders using data from the autism and developmental disabilities monitoring network. *Environmental Health Perspectives.* April 17. Epub.

Kanner L. 1943. Autistic disturbances of affective contact. *Nervous Child.* 2:217–50.

Karlsson H et al. 2012. Maternal antibodies to dietary antigens and risk for nonaffective psychosis in offspring. *The American Journal of Psychiatry.* June;169(6):625–32.

Kharrazian D. 2010. *Why Do I Still Have Thyroid Symptoms? When My Blood Tests Are Normal: A Revolutionary Breakthrough in Understanding Hasimoto's Disease and Hypothyroidism.* Morgan James Publishing.

Kim YS et al. 2011. Prevalence of autism spectrum disorders in a total population sample. *The American Journal of Psychiatry.* May 9; online.

King MD, Bearman PS. 2011. Socioeconomic status and the increased prevalence of autism in California. *American Sociology Review.* April 1;76(2): 320–46.

Koolhaas JM et al. 2011. Stress revisited: a critical evaluation of the stress concept. *Neuroscience Behavioral Review.* April;35(5):1291–301.

Krakowiak P et al. 2012. Maternal metabolic conditions and risk for autism and other neurodevelopmental disorders. *Pediatrics.* May;129(5): e1121–28.

Land RJ et al. 2012. Latent class analysis of early developmental trajectory in baby siblings of children with autism. *Journal of Child Psychology and Psychiatry.* May 10. Epub ahead of print.

Landrigan P, Lambertini L, Birnbaum L. 2012. A research strategy to discover the environmental causes of autism and neurodevelopmental disabilities. *Environmental Health Perspectives.* April 25. Epub ahead of print.

Lazarev VV et al. 2010. Interhemispheric asymmetry in EEG photic driving coherence in childhood autism. *Clinical Neurophysiology.* Feb.;121(2): 145–52.

Leisman G, Melillo R. 2009. EEG coherence measures functional disconnectivities in autism. *Acta Paediatrica*. 98(460):28–29.

Luby JL et al. 2012. Maternal support in early childhood predicts larger hippocampal volumes at school age. *Proceedings of the National Academy of Sciences of the United States of America*. Feb. 21;109(8):2854–59.

Lundstrom S et al. 2007. Autism spectrum disorders and autistic-like traits: similar etiology in the extreme end and the normal variation. *Archives of General Psychiatry*. Jan.;69(1):46–52.

Lyall K et al. 2012. Pregnancy complications and obstetric suboptimality in association with autism spectrum disorders in children of the Nurses' Health Study 11. *Autism Research*. Feb.;5(1):21–30.

Malik M et al. 2011. Expression of inflammatory cytokines, Bcl12 and cathepsin D are altered in lymphoblasts of autistic subjects. *Immunobiology*. Jan.–Feb.;216(1–2):80–85.

Mann JR et al. 2010. Pre-eclampsia, birth weight, and autism spectrum disorders. *Journal of Autism and Developmental Disorders*. May;40(5):548–54.

McGilchrist I. 2011. The divided brain. www.ted.com/talks/iain_mcgilchrist_the_divided_brain.html.

McGilchrist I. 2010. *The Master and His Emissary: The Divided Brain and the Making of the Western World*. Yale University Press.

Melillo R. 2009. *Disconnected Kids*. Perigee.

Melillo R, Leisman G. 2009. Autistic spectrum disorders as functional disconnection syndrome. *Review of Neuroscience*. 20(2):111–31.

Melillo R, Leisman G. 2004. *Neurobehavioral Disorders of Childhood: An Evolutionary Perspective*. Springer Science+Business Media.

Miles JH et al. 2005. Essential versus complex autism: definition of fundamental prognostic subtypes. *American Journal of Medical Genetics*. June;135(2):171–80.

Millin PM et al. 2011. Prenatal exposure to hyperemesis gravidaram linked to increased risk of psychological and behavioral disorders in adulthood. *Journal of Developmental Origins of Health and Disease*. Aug. 10; online.

Miodovnik A. 2011. Environmental neurotoxicants and the developing brain. *The Mount Sinai Journal of Medicine*. Jan.–Feb.;78(1):58–77.

Parner ET et al. 2012. Parental age and autism spectrum disorders. *Annals of Epidemiology.* Jan. 24; online.

Paul LK. 2011. Developmental malformation of the corpus callosum: a review of typical callosal development and examples of developmental disorders with callosal involvement. *Journal of Neurodevelopmental Disorders.* March;3(1):3–27.

Pembrey ME. 2002. Time to take epigenetic inheritance seriously. *European Journal of Human Genetics.* Nov.;10(11):669–71.

Remington AM, Swettenham JG, Lavie N. 2012. Lightening the load: Perceptual load impairs visual detection in typical adults but not in autism. *Journal of Abnormal Psychology.* May;121(2):544–51.

Reynolds RM et al. 2007. Stress responsiveness in adult life: influence of mother's diet in late pregnancy. *The Journal of Clinical Endocrinology & Metabolism.* June;92(6):2208–10.

Robinson EB et al. 2011. Evidence that autistic traits show the same etiology in the general population and at the quantitative extremes (5%, 2.5%, and 1%). *Archives of General Psychiatry.* Nov.;68(11):1113–21.

Roelfsema MT et al. 2011. Are autism spectrum conditions more prevalent in an information-technology region? A school-based study of three regions in the Netherlands. *Journal of Autism and Developmental Disorders.* June 17; online.

Rubio-Tapia A. et al. 2009. Increased prevalence and mortality in undiagnosed celiac disease. *Gastroenterology.* July;137(1):88–93.

Rutter M. 2005. Incidence of autism spectrum disorders: changes over time and their meaning. *Acta Paediatria.* Jan.;(1):2–15.

Schipul SE et al. 2012. Distinctive neural processes during learning in autism. *Cerebral Cortex.* April;22(4):937–50.

Schmidt RJ et al. 2012. Maternal periconceptional folic acid intake and risk of autism spectrum disorders and developmental delay in the CHARGE (CHildhood Autism Risks from Genetics and Environment) case-control study. *American Journal of Clinical Nutrition.* July;96(1):80–89.

Schmidt RJ et al. 2011. Prenatal vitamins, one-carbon metabolism gene variants, and risk for autism. *Epidemiology.* July;22(4):476–85.

Silberman S. 2001. The geek syndrome. *Wired*. www.wired.com/wired/archive/9.12/aspergers.html.

Steffenburg S et al. 1989. A twin study of autism in Denmark, Finland, Iceland, Norway and Sweden. *Journal of Child Psychology and Psychiatry*. May;30(3):405–16.

Van Dijk SJ. 2009. A saturated fatty acid-rich diet induces an obesity-linked proinflammatory gene expression profile in adipose tissue of subjects with metabolic syndrome. *The American Journal of Clinical Nutrition*. Dec.;90(6):1656–64.

Vargas DL. 2005. Neuroglial activation and neuroinflammation in the brain of patients with autism. *Annals of Neurology*. Jan.;57(1):67–81.

Waterland RA, Jirtle RL. 2003. Transposable elements: targets for early nutritional effects on epigenetic gene regulation. Aug.;23(15):5293–5300.

Wicker B et al. 2008. Abnormal cerebral effective connectivity during explicit emotion processing in adults with autism spectrum disorder. *Social Cognitive and Affective Neuroscience*. June;3(2):135–43.

Williams JG, Higgins JPY, Brayne CEG. 2006. Systematic review of prevalence of autism spectrum disorders. *Archives of Disease in Childhood*. 91:8–15.

Wolff S, Narayan S, Moyes B. 1988. Personality characteristics of parents of autistic children: a controlled study. *Journal of Child Psychology and Psychiatry*. March;29(2):143–543.

Zerbo O et al. 2012. Is maternal influenza or fever during prepregnancy associated with autism or developmental delays? Results from the CHARGE (CHildhood Autism Risks from Genetics and Environment) Study. *Journal of Autism and Developmental Disorders*. May 5. Epub ahead of print.

Zerbo O et al. 2011. Month of conception and risk of autism. *Epidemiology*. July;22(4):469–75.

For a full list of the author's presentations and publications, please visit www.BrainBalanceCenters.com.

INDEX

Page numbers in **bold** indicate charts or tables; those in *italic* indicate illustrations.

Dr. Robert Melillo has been an active clinician for more than twenty-five years, and he is one of the world's most respected specialists in childhood neurological disorders. His areas of expertise include autism spectrum disorders, ADD and ADHD, dyslexia, Asperger's syndrome, Tourette's syndrome, bipolar disorder, and other attention, behavioral, and learning disorders. He also specializes in neuroimmune disorders in children and adolescents, such as PANDAS and PANS. Dr. Melillo also focuses on preconception programs for adults. He has been working with children with neurological disabilities for more than twenty years.

His work as a university professor and a cutting-edge brain researcher and his success with more than a thousand children in his private program led to the creation of Brain Balance Achievement Centers. The centers use a medicine-free, multimodal curriculum focused on correcting the primary issue inherent in most neurological developmental disorders—a functional disconnection in the brain in which one side is growing too slow or too fast, preventing the two sides from integrating and working in harmony.

Dr. Melillo and his research partner, Dr. Gerry Leisman, are considered two of the world's leading experts and pioneers in functional disconnection and its relationship to neurobehavioral disorders. Since they introduced the concept, functional disconnection has become the leading theory as the key issue related to autism, ADHD, dyslexia, and other similar conditions. Their work is leading the way toward understanding the underlying nature of these disorders and their causes. Their lab is focused on developing effective treatments.

Dr. Melillo wrote a working-theory textbook on development disabilities called *Neurobehavioral Disorders of Childhood: An Evolutionary Perspective*, which was published in 2004. In 2009, Perigee published his first book for a lay audience called *Disconnected Kids*. Greeted with an overwhelmingly positive response, the book serves as the foundation for the work being done at the one hundred Brain Balance Achievement Centers around the country. His second book, *Reconnected Kids*, was published by Perigee in 2011.

Dr. Melillo's leadership in the field, as well as his personal dedication to helping children with neurological disorders, has made him one of the most sought-after speakers on the subject, both in the United States and abroad. His optimistic and straightforward approach to overcoming childhood brain disabilities has given hope to countless families throughout the world.

Dr. Melillo has made dozens of national and local television appearances and has been interviewed on hundreds of radio programs. He frequently appears as an expert commentator on Fox News and NBC.

Dr. Melillo is an affiliate professor of rehabilitation sciences at Nazareth Academic Institute and a senior research fellow with the National Institute for Brain and Rehabilitation Sciences. He is a postgraduate professor of childhood developmental disabilities. He holds a master's degree in neuroscience, a master's degree in clinical rehabilitation neuropsychology, and is completing his doctorate in the same subjects. He holds a doctorate in chiropractic, a diplomate in neurology, Fellowship American College of Functional Neurology, Fellowship American Board Childhood Developmental Disabilities, and is the executive director of the F. R. Carrick Research Institute and the Children's Autism Hope Project. He is also president of the International Association of Functional Neurology and Rehabilitation and the coeditor in chief of the professional journal *Functional Neurology, Rehabilitation and Ergonomics*. He has published numerous scientific papers and contributed chapters to seven professional books. He has also made hundreds of conference presentations.

Dr. Melillo maintains a private practice in New York and in Atlanta. He lives in Rockville Centre, New York, with his wife and three children.